THE SNAKEBITE SURVIVORS' CLUB

JEREMY SEAL

THE SNAKEBITE

SURVIVORS' CLUB

TRAVELS AMONG SERPENTS

A HARVEST BOOK
HARCOURT, INC.
San Diego   New York   London

www.harcourt.com

First published in England in 1999 by Picador

Quotes from *Foot-Loose in India* by Gordon Sinclair,
reprinted by permission of Henry Holt and Company,
LLC.

The quotation from *Travels in the Coastlands of British
East Africa*, by William Walter Augustine Fitzgerald,
published by Chapman & Hall, 1898, is used with kind
permission from Kluwer Academic Publishers.

Library of Congress Cataloging-in-Publication Data
Seal, Jeremy.
      The snakebite survivors' club: travels among
      serpents / Jeremy Seal.—1st ed.
         p. cm.
      Includes bibliographical references (p.   ).
      ISBN 0-15-100535-4
      ISBN 0-15-601367-3 (pbk.)
         1. Poisonous snakes—Anecdotes. 2. Snakebites—
      Anecdotes. 3. Seal, Jeremy—Journeys.
      I. Title.

QL666.O6 S3917 1999
597.96'165—dc21   99-049713

Text set in Fairfield Medium
Book designed by Kaelin Chappell
Printed in the United States of America
First Harvest edition 2001
J I H G F E D C B A

To my darling mum, with love

CONTENTS

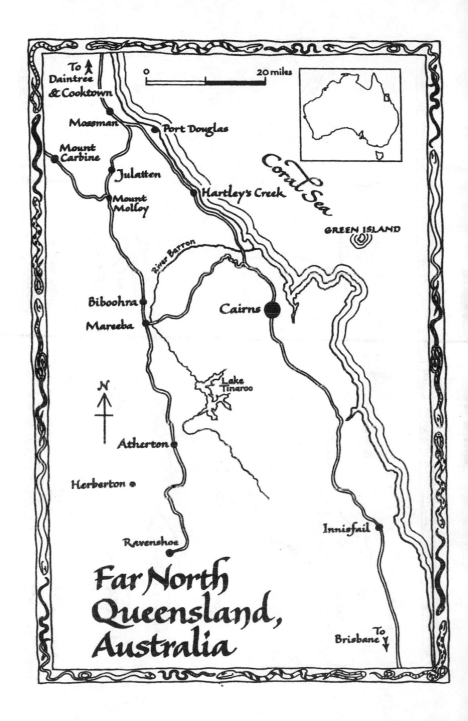

To Daintree & Cooktown

20 miles

Mossman

Port Douglas

Mount Carbine

Julatten

Coral Sea

Mount Molloy

Hartley's Creek

GREEN ISLAND

River Barron

Biboohra

Cairns

Mareeba

N

Lake Tinaroo

Atherton

Herberton

Innisfail

Ravenshoe

Far North Queensland, Australia

To Brisbane

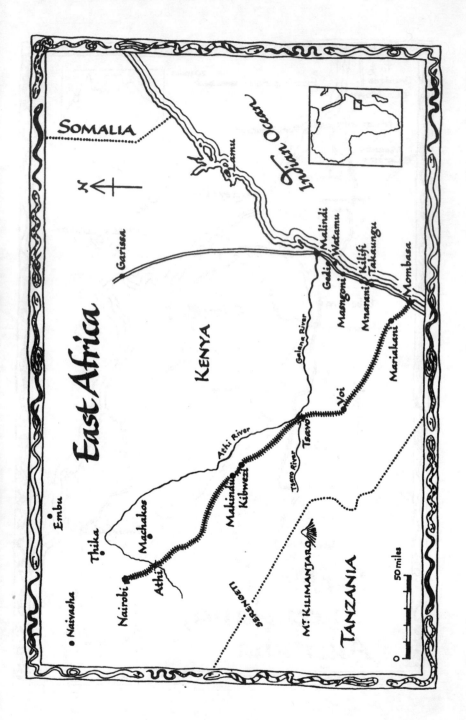

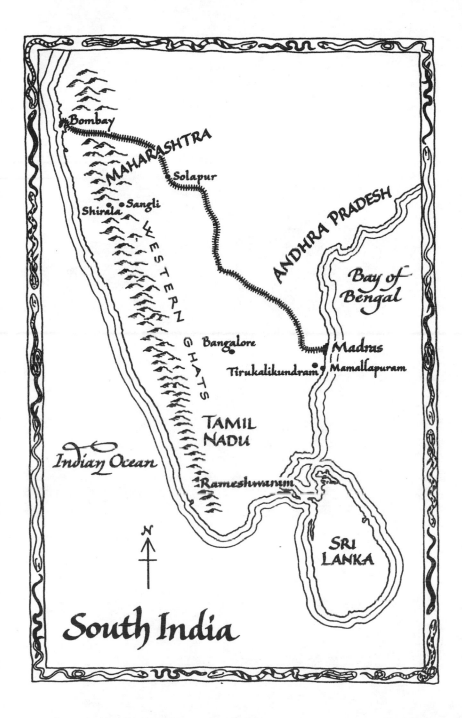

Bombay

MAHARASHTRA

Solapur

Sangli
Shirala

WESTERN GHATS

ANDHRA PRADESH

Bay of
Bengal

Bangalore

Madras

Tirukalikundram · Mamallapuram

TAMIL
NADU

Indian Ocean

Rameshwarum

N

SRI
LANKA

South India

PROLOGUE

I stood in the doorway, deferring the moment with a fretful cigarette. One of the gorillas across the way returned my gaze
through the bars of its cage, and we eyed each other for some
time. The ape had sad, fathomless eyes. The human, I guessed,
must have anxious, haunted ones.

The gorilla scratched itself, yawned and toppled back into
the straw. I turned towards the building behind me. It was
square and neo-classical in style. A decorative frieze ran round
the portals in which were carved iguanas, snails, salamanders,
frogs, alligators, pythons and, in place of prominence on the
keystone, three hooded cobras wearing berets of black bitumen
that had dripped from the roof in the heat of a former London
summer. A group of school children approached. "Oh no,
Priscilla!" yelled a young boy. The contents of the building had
dawned upon him. "This is the scary one!"

Years before, as a child, I had stood in this same doorway. I
remember how I had imagined within a bottomless pit overflowing with snakes, plaits of snakes spilling on to the shiny
floor to untangle themselves and slither numberless down long
corridors, and the images had stopped me in my tracks. My
palms had prickled with sweat and tears had welled up in my
eyes, and thirty years later little had changed.

I was not, of course, alone in my fear of snakes. Ever since
the world's first "serpentarium" or "reptilium" opened at
London Zoo in June 1849, in the Swiss chalet-style building
which had formerly housed the zoo's carnivores, children have
screamed between the portals and adults have turned away,
pale-faced. Beatrice Thompson, who enjoyed frequent visits to
the zoo at the end of the last century, remarked how she had

"known many women of the educated classes firmly refuse to enter the Reptile House at all."

True, the image of the overflowing snake-pit had since left me, but a more particular, more threatening vision had taken its place: a single snake seething at my feet, a wave of chevrons and diamonds breaking yellow and brown, then the instant of sharpness and blood nudging scarlet from the puncture marks as the disappearing snake flicked its tail at the grass. The snakebite sequence began as a recurring dream. I then started imagining it, imagined it so often that I began to anticipate it and it became in time a fact of my future, as if I was living in a world littered with live ordnance which I was one day destined to tread upon. I began to believe that I was going to die of snakebite.

So what was I doing here, so close to my fear? The answer perhaps lay in an 1805 penny tract called *Wonderful Account of the Rattlesnake and Other Serpents of America* that I recently came across in a library, where I found the suggestive description of snakes possessing their prey—birds, rabbits and other rodents—"with infatuation," so that they "flutter or move slowly, but reluctantly, towards the yawning jaws of their devourers, and creep into their mouths..." I knew that feeling. So too, I suspect, had everyone who had ever stood at the entrance of reptile houses and snake parks, daring themselves to cross the threshold despite their abhorrence of what they would find inside. I was drawn to snakes even as they disgusted me. The fascination was as strong as the fear; a kind of transfixing awe, an infatuation, was the result. The presence of snakes seemed to cast a strange light across the landscape, a light that might attend the end of the world: spare, stark and alive with a crackly energy. A charge pulsed through such moments, and made them among the most compelling of my life.

Just as I could not keep myself from snakes, so I was drawn to survivors of snakebite, that singular clique, that club whose members had been touched by something dark, exotic and preternatural. They were like those who had experienced ship-wreck, lightning strike or even alien abduction; they wore a badge of otherness that the rest of us could only shiver at even as we asked, our mouths slack with fascination: *being bitten; what did it feel like?* That was the point; we wanted to know. If we examined our curiosity deep enough, I suspected we might even acknowledge that these people, these survivors, had experienced something far stranger than we would ever know, and what we truly felt, despite our shivers, was something akin to a kind of dangerous envy, even a yearning for the experience they had survived.

Which perhaps explained why I should be standing outside the reptile house at London Zoo on a spring afternoon in 1996, gathering myself to step inside.

1. THE REPTILE HOUSE

I do not think you will spend a very long time in the Reptile House at the Zoo.

—Enid Blyton, *The Zoo Book*

MARCH 1996

I eventually stepped inside, of course. I'd eventually stepped inside back in my childhood, just as many others had done before. The contents of the reptile house finally drew us in. E. G. Boulenger, the zoo's aquarium director in the 1920s, called the entrance to the reptile house, and the conflict between fear and fascination that habitually played there, the site of the zoo's one truly "interesting psychological entertainment." The snake show, which testified to man's fascination for these creatures, had been around in its various guises long before the opening of the reptile house. In the seventeenth century, halfpenny promoters and impresarios discovered how the terrible magnetism of boa constrictors and other exotic species could transform the appeal of their travelling menageries. Those specimens that survived the uncertain sea crossing home from Brazil, the Far East or the American colonies, were billed as fabulous, fatal creatures from distant lands, and attracted rapt audiences when they were displayed in London rooms. A BEAUTIFUL RATTLESNAKE ALIVE, read a headline in the *General Advertiser* in January 1752.

This exotic Animal is extremely well worthy the Observation of the Curious; Its Eyes are of great Lustre, even equal to that of a Diamond, and its Skin so exquisitely mottled and of such surpassing Beauty as baffles the Art of the most celebrated Painter...There is not the least cause for Fear, though it were at Liberty in the Room: but that the Ladies may be under no Apprehension on that Account, it is kept in a Glass-Cage. It is very Active, and is the first ever shown alive in England.

The zoo's first snakes were mostly from the Royal Collection, which had formerly been housed in the Tower of London, and were presented to the Royal Zoological Society, the zoo's patron, by King William in 1831. Lack of suitable housing consigned the collection, which included an Indian python, a huge constricting anaconda and over a hundred rattlesnakes, to spending its first eighteen years at the zoo in cramped boxes. The opening of the reptile house at least allowed the survivors—there were some twenty snakes on show at the beginning—space to move and branches on which to arrange themselves, even if the pythons tended to eat the blankets that were provided to supply warmth during the winter.

Crowds flocked to the zoo in their top hats and bonnets. Visitors doubled to over 300,000 during 1850, but the public enthusiasm was not entirely due to the snakes. The arrival of a hippo in late May of that year, the first one seen live in Europe since Roman times, also aroused enormous curiosity. Among the retinue that had collected around the celebrated hippo in the course of its long journey from Africa to Regent's Park were two Egyptian snake charmers, who agreed to give a display at the reptile house the day after their arrival. A naturalist called Broderip was present at the display, and remembered how "there was not much difficulty in getting a front place, but those behind pressed the bolder spectators rather inconveniently forward." Snake shows tended to draw impressive crowds, but front-row seats were guaranteed available. The same paradoxical feeling I'd known in the doorway, the irresistible allure and dread that snakes simultaneously inspired, as if the audience needed to be there, just not *right* there, was evident even in the manner in which the snake-show crowd arranged itself.

As the crowd looked on, the senior charmer commanded his assistant to remove an Egyptian cobra from "a large deal box with a sliding cover." "The serpent," wrote Broderip, "instantly raised itself, expanded its hood, and turned slowly on its own axis, following the eye of the young Arab, turning as his head, or eye, or body turned." Later, the older Arab brought his face close to that of the raised cobra, daring it to strike. "Suddenly," wrote Broderip, "it darted open-mouthed at his face, furiously dashing its expanded whitish-edged jaws into the dark hollow cheek of the charmer..." That both charmers took apparent bites from cobras that day, and suffered no ill effects, convinced the scientist in Broderip that the fangs had been removed. Not that he retreated any less readily than his stampeding fellow spectators when one of the snakes temporarily escaped the reach of the charmers.

That the Egyptians came to no harm that day may have inspired reptile-house attendant John Girling's tragic *folie de grandeur* two years later. On the morning of October 20, 1852, after seeing off a friend bound for Australia which had occasioned a night's determined carousing consisting, according to *The Times*, of three pints of beer, "a quartern of gin at the public-house in Shoe-Lane, another afterwards, and again another at 8 o'clock" in the morning, Girling felt unwisely moved to demonstrate his own snake charming skills. Swaying uncertainly, he removed a "Morocco" snake from its cage and draped it around the reluctant neck of a passing attendant from the humming-bird collection, before returning it to its cage and proclaiming, "Now for the cobra," which he grabbed and proceeded to swing around his head. This cobra, whose fangs were perfectly intact, promptly struck Girling on the nose. He was immediately conveyed to the University Hospital in Gower

Street, where he died shortly after nine o'clock, less than an hour after he was bitten.

In the aftermath of Girling's death, *The Times* was quick to reassure the general public about safety standards at the reptile house. "Visitors," it explained, "are enabled to see the serpents in perfect security, through the thick glass fronts of the compartments, and nothing can be better than the arrangements of the Society in this portion of their display..."

So, I was safe from snakebite. I was in a wide, soft-lit corridor that returned to its beginning, describing a rectangular tour around a large, central island. The only light issued from the glass-fronted pens sunk flush into the walls on either side of me. They resembled a series of screens broadcasting artificially floodlit landscapes; montages of rock and branch, and scattered brown leaves dry as biscuits in which little activity was evident. I steadied myself; I was among snakes.

Children raced between the pens, recharging their excitement at each visit. The adults moved with a whispered, library calm, unaware that they were among the deaf. All snakes can detect are ground-borne vibrations, a fact I'd long since absorbed if only for self-preservation's sake. That snakes spoke the language of vibration may have been the limit of my herpetological knowledge, but I at least exploited the fact to the full whenever I was in even the most mildly suggestive surroundings. As for my personal minefield, long grass, I would stamp my way through, giving snakes every warning of my approach and relying on their reported readiness to flee. In fact, as the puzzled stare of a stranger informed me, the proximity of snakes had evidently triggered the same response at the reptile house. I was unconsciously clomping down the corridor, like some seasonally confused loon thinking to kick the snow off his boots. Sheepish, I apologized and strove to proceed more normally.

I had a particular purpose at the reptile house. I was not here merely to reacquaint myself with surroundings that transfixed me just as much as they appalled me. I was looking for particular snakes. Over the years, as if to increase their hold upon me, my feelings of fear and fascination for these creatures had seemed to batten upon four particular species of snake; snakes that had shed their herpetological confines to become creatures of fable and superstition, laden with meaning, symbol and allure, but capable of extraordinary virulence too; snakes that drew me towards them like night creatures which I was powerless to resist.

I found them in the reptile house's central island. The snakes exhibited here were inmates in a kind of reptilian maximum-security wing, self-enclosed so any escapees presented no threat to the public. This, then, was where they kept the highly venomous species. Until recently, the spirit of that memorably melodramatic Victorian phrase, the "thanatophidia" or the "death-dealing snakes," had been echoed by the sign introducing this section of the reptile house as "The Killers."

And so I came to the first of my four snakes. The Indian cobra lay hidden among foliage and only a brief portion of its body was visible, sleek as a new tire. Even so, what I could see of the real thing seemed to have no connection with the hooded posture that has come to be this snake's signature, a byword for a kind of absolute intent that has persuaded manufacturers of everything from cars to camera equipment, helicopters to beer, shoes to overhead projectors, golf clubs to radios and surfboards to stroboscopes, as well as millions of devotional Hindus in modern India, to entrust their products or even themselves to its dark, iconic allure.

As the display on the pen explained, the snake that had accounted for Girling still killed between 10,000 and 20,000 people per year in India. In Girling's day the Indian cobra was the most feared of all snakes and attracted melodramatic epithets like "grim death" and "awful presence." It was also considered irascible. In the 1870s the lower glass sections of the zoo's cobra pens were painted over to stop the snakes striking at passing humans, and damaging themselves against the panes in the process.

I walked on, for if the cobra had been accounted the most notorious snake of all during the nineteenth century, then another species entirely could lay claim to that title in our century. The black mamba of Africa, said the display, was "one of the world's most deadly snakes; one that came to the zoo in 1954 was supposed to have killed six people." This particular snake, however, was in no mood for killing. As I watched, it began to shed its six-foot skin. By rubbing its head against a branch, it forced a tear in the old skin. Now the shedding could begin. The snake inched itself clear with a series of patient, tiny shrugs, as if from a tight sleeping-bag that it turned progressively inside out. The old skin peeled away, trailing from the snake, desiccated, dandruffed but intact, to reveal a skin beneath that was shiny new, almost laminated.

The mamba had surprised me. I had imagined myself meditating on the death-dealing malevolence of this snake. Instead, as the mamba rid itself of the old skin with a final thrust of its tail half an hour later, I realized I'd been watching a creature return itself to an ageless, pristine beauty. As I turned away, I noticed a crowd had gathered around the pen, transfixed by the rebirth of this black mamba. Outside, lights were coming on and the air was full of faxes and radio waves and passenger jets, and we were reduced to silence at the heart of the city as this creature stripped itself of its every blemish, and returned to a new beginning that we could only dream of, mocking our mortality. I'd just had a glimpse of the snake's other self, its capacity for magic as well as for murder.

Less widely known but accounted more deadly yet was my third snake, the taipan of Australia which, the display explained, "produces venom that is six times more deadly than that of the Indian cobra." Before the availability of antivenom (the antibodies given intravenously that counter the effects of venom), only one person was known to have survived the bite of a taipan, a nineteen-year-old Queensland man in excellent physical condition who was bitten through the highly mitigating effects of boot, thick socks and calloused skin. I had been in rapt awareness of the taipan ever since the otherwise dryish Hamlyn all-colour paperback *Snakes of the World* hyperbolically informed me in my early teens, when my dreaded fascination was well established, that "if a man is bitten by this monster there is little hope of recovery and death within a few minutes is the most likely result." The rust-red taipans were compellingly active, their slender five-foot bodies fashioning momentary cat's cradles among the branches in their pen.

It was proving an instructive visit. I was beginning to appreciate what I was doing here, coming back to this place that had haunted me since my childhood. It was a way of gently acquainting myself with the snakes I most feared, a chance to view them from the safe side of thick glass before the rather more serious business ahead. Soon, I would set out to confront these snakes on their own ground in India, Africa, Australia, and, the habitat of my fourth and final snake, America. I was not yet sure what I hoped to achieve in those distant places, except that I might come to understand something of the remarkable, sometimes paralyzing effect these creatures seemed to exert upon mankind and, more particularly, upon myself. Whether I wanted to make these journeys or whether I was being drawn to make them despite myself I could not say. Such was the singular effect of snakes.

Snakes in the wild, of course, were a different prospect altogether to these captive specimens and I wondered whether I was up to the journeys I had planned. But whenever the fear came upon me, I could always summon up the snakebite survivors' club, a club that met only in my mind to intrigue me, but also to give me solace. For if snakebite was my greatest fear, then snakebite survivors were the psychological antidote, living evidence that snakebite need not kill. I thought of snakebite survivors just as, I liked to think, racing drivers bolstered their nerves by recalling the miraculous escapes from high-speed crashes of their colleagues or skin divers remembered incredible stories of survival from shark attack; that if *it* happened, there was still hope. Plenty of people survived snakebite: the words ran round my mind until they became a mantra, a reassurance that I could survive my experience of snakes.

And so I came to my fourth snake, the American one, and leaned my hand against the glass as if to touch it. Which was when the rattlesnake seemed to explode; its body a writhing ribbon of diamonds uncoiling like a harpoon rope behind the heavy, bulbed head. Its mouth yawed open in a blurry, pink-mouthed strike. It hit the glass where my left hand lay.

2. AMERICA I

DEFENSE COUNSEL: How many of you are afraid of snakes?
(*Many jurors raise hands.*)
DEFENSE COUNSEL: How many of you are not afraid of snakes?
(*No response.*)

   —Trial transcript extract, *The State* versus *Summerford*, February 1992,
   Jackson County Courthouse, Scottsboro, Alabama

APRIL 1996

On a Friday evening in the fall of 1991, as people all over
Alabama sat down to TV dinners, or mowed the twilit lawn for
the last time till spring, Darlene Summerford Collins took a
bite from a diamondback rattlesnake. She saw only a blur of
motion, a pink flash of gaping jaws as it lashed up at her de-
scending hand. Its fangs found flesh high on the thumb, close
to the wrist joint of her left hand.

Darlene had been bitten before. The Lord knows, she'd
handled enough snakes over the years. Used even to carry photos of snakes in her purse to show to friends; favourite ones—
copperheads, moccasins, that greenish timber rattler she so
liked. And her husband, Glenn, had been handling them for the
best part of ten years, lately in the little church with the teetering steeple on Woods Cove Road outside of town, letting them
get into his hair or using them to wipe the sweat from his face,
handling Satan just like any true believer.

Only now, her husband was standing over her with a gun, the
pistol she'd bought at Scottsboro Pawn just six months earlier,
his hand twisting cruel, thick plaits in her red hair as she

looked down at her thumb and the bead of blood that stood proud a brief moment before running across her palm, seeking out the lifelines.

"The lid," hissed Glenn. "Put the lid back on the box." As she did so, the snake rattled its tail in anger. She had heard its tindery buzz many times, but never before had the sound made her nauseous. In the musty shed, she could smell the snake and the fetid scent of its anger, the drink on her husband's breath and the coal-dust smell of gunmetal. Now Glenn was dragging her to another box. She could plainly see the contents: two large rattlesnakes whose diamond-patterned backs were faintly echoed in the mesh tracery of the chicken-wire lid.

"You praying now?" he whispered. He smiled. "I bet you praying now."

Foreboding coiled tight in her belly. Around her thumb, there was already a vicious clamping pain where the venom was going to work. But what was making her sick was something else: it felt as if her senses had gone. She was sure she could hear the river—and when had she ever done that?—and smell doughnuts from somewhere, and every consonant Glenn had uttered was now careening around her head, just in no particular order and a couple of octaves lower. There were times she had thought every moment in her life was leading her downwards, but not to this, not to murder from multiple rattlesnake bites. Not by her husband.

~

That was over four years ago. By the time I reached the house, Glenn and Darlene were long gone, but something of that distant fall evening still clung to the place, a baleful, poisoning mood that had spited time and made me shiver. The house lay beyond Scottsboro's outskirts, where the neat lines of weatherboard homes fronted by mailboxes gave way to the scruffy smallholdings on Barbee Lane, a long nibbled ribbon of blacktop that cut into the countryside. Chained dogs lay among the

gutted auto parts—the engine blocks, radiators and rusted ex-
hausts—of pick-ups spray-painted in bright pinks and purples.
Workshop radios interspersed music and commercial messages.
Save a light breeze mussing its mane, a single horse stood motionless in a field.

The house stood at Barbee Lane's far extent, among the
stands of pine, hickory and beech that led down to the
Tennessee River. The river threw a big southerly lasso loop
through Alabama. It entered the state in the north-east from
Tennessee and, mindful of its name, left it for Tennessee to the
north-west. For much of its length, however, it was a shapeless
thing. A series of bloated blackwater sloughs had backed up behind dams all the way to the Mississippi, depriving the river of
its sinuous, serpentine coils. As if in compensation for the loss
of its grace, the flabby river had at least acquired an excellent
reputation for catfish, lurking monsters that weighed up to a
hundred pounds. Fishermen never knew what might fetch up
along the Scottsboro stretch of the Tennessee.

The Barbee Lane house lay in permanent shadow. It had a
tin roof, ripped tar-paper walls and a highly visible sign warning
against trespass. A cardboard-coloured dog, a large one, barked
fiercely and balked my uncertain approach. For a while, we just
looked at each other. The dog's look was more direct. Guessing
he might not be a patch on the house's current tenant, I returned to the car.

I drove past the imposing brick courthouse in Scottsboro's
main square, and the flourishing attorneys' offices that fed off
it. Beyond their brief affluence, however, were autopart stores
emphasizing the essential—"Need a new engine? We can
help"—and pawn shops that had long since quit claiming any
measure of discernment—"Car title pawn. And guns, jewellery,
anything"—and the unclaimed luggage dealers that airports
from New Orleans to Atlanta supplied with job lots of umbrellas, coats, suitcases, boots, hats, paperbacks, bottles of aspirin,
Walkmans, laptops, pens, even forgotten photos of loved ones

caught in distant, suburban sunshine, posing beside car ports in Ohio, or on the balconies of Florida condos with just-about views of the ocean.

Woods Cove Road wound out of town on the west-side, passed the Jackson Country hospital, crossed the railroad tracks and headed through cornfields before being sucked into the wake of the highway to die premature, serving just a few loose farmsteads—and the old church—on the way. The church looked old enough. Less like a church though. Concrete plinths where the fuel pumps once stood still fronted the former country store and gas station, set to catch unwary shins. The building was flat-roofed, whitewashed and jerry-built. Spare squares of sorry chipboard patched the front. A poor, hand-painted sign said "Paster Frank. Everyone welcome. Friday night. Sunday night. 7:00 p.m." Pastor Frank, I figured, must have followed in Glenn's footsteps for a while. The first rime of rust on the padlock suggested, at best, an irregular ministry.

I rubbed at the window. A spider's web slung across the pane gave even the indistinct gloom within a fractured appearance. There was the suggestion of old linoleum switch-backing across the floor. A chair lay on its back, like a man with his knees tucked up to his chest as if to protect himself from an unholy kicking. Around the back of the church, a stack of decrepit pews were turning into mold among the sycamore trees, their ply sheets yawning open like the pages of well-thumbed Bibles.

More effort had gone into painting the church's name, The Church of the Lord Jesus Christ, in a blue arc around a cross. In Glenn's time, I mused, this had been the Church of Jesus With Signs Following. At that time the words "Mark 16, vs 15 18" were painted with a shaky reverence on an inside wall; "And these signs shall follow them that believe," read the verses referred to. "In my name shall they cast out devils; they shall speak with new tongues; They shall take up serpents; and if they drink any deadly thing, it shall not hurt them; they shall lay hands on the sick, and they shall recover." In Glenn's church, like others scattered throughout the Southern Appalachian mountains, a place unto itself among a jumble of intersecting southern state lines, they had taken the Bible at its every word.

Glenn Summerford did the lot. Spoke in tongues, drank deadly strychnine and drain cleaner, and even took electric shocks in God's name. But it was the snake-handling he really took to. Did it with an abandon that frankly scared his packed congregation as much as it inspired them. Glenn had come to the Lord the hard way, leaving in his turbulent wake a trail of criminal charges from illegal racoon-hunting and grand larceny to the attempted murder of his mother-in-law—with a vase.

Darlene could tell you all about his violent temper, drinking bouts and infidelities. She was one of thirteen children who had grown up on handouts near Dutton township on Sand Mountain, the looming plateau on the east-side of the Tennessee River that stands between Scottsboro and the sunrise. She and Glenn had met when they were young. By the mid-1970s, she was falling for his considerable, if diabolical charisma. He had fire and life. What he'd noticed were her enormous blue eyes and that hair you couldn't ignore, ginger-red as agricultural twine. Within months of their marriage, however, he was kicking her around, accusing her of cheating on him with young boys, even with her own brother, and drinking with a ferocity that drove him still deeper into sloughs of jealousy and murderous rage.

Then, one day in 1982 Glenn wove out of a bar and, promising himself a new start, fell into a friend's pick-up bound for a church service across the state line in Kingston, Georgia. Propped in a pew, he watched men remove rattlesnakes from wooden boxes stacked by the altar, and pass them among themselves during the singing with a strength and confidence that he had only ever got out of bottles. He could not hear the snakes' angry rattles above the cacophonous wailing, the Hallelujahs and the music—thumping drums, twanging guitars, crashing cymbals and the roar of miked-up male voices—but he could see them tense in the initial grip of their handlers and then, miraculously, go rope-limp. Even as the men draped them round their shoulders or held them by the midriff, letting them hang by their sides, the snakes made no attempt to strike. One man tucked a young rattler in his shirt pocket; its tail protruded like a limp, gorgeously patterned dandy's kerchief. Later, when things got a little crazy—women were keeling over, shuddering like cardiac cases or were slumped, sobbing against the altar— the men even threw the rattlesnakes into the air, sending light-bulbs swinging, hustling the shadows away, and caught them on their descent.

It might, granted, have been on the road to Cartersville, the nondescript strip of fast-food joints and budget motels huddled along Georgia Highway 41 some forty miles north of Atlanta, rather than to Damascus, but Glenn nevertheless sensed re-demption in that small, clapboard church on a hill. This was simple instruction. The Word of the Book guaranteed salvation through the act of lifting the hinged wire lid of a wooden box, reaching in and taking from it writhing rosaries of evil incar-nate. Belief would do the rest. Faith would disarm the devil. Perfect victory, they called it. By the end of the service, Glenn was converted. He was set on becoming a preacher, and handling snakes.

Soon, he had seventeen snakes in the shed at the bottom of Barbee Lane. Supportive preachers provided him with some. Others told him to go out into the woods, to sit down and pray and let the Lord send the snakes to him. Some he bought for a few dollars from local boys who caught them in the brushwood. He even found a couple of his own, sunning themselves among the pine trees in the spring and learned how to box them with-out harming himself. He took them to his first church services in Scottsboro, itinerant, short-notice arrangements that con-vened in borrowed rooms and old buildings all over town. His congregation even did a spell above a chicken restaurant.

Locals accused them of being "freaks," "Holy Rollers" and "Jesus Onlys," even broke what panes remained in their windows. So they came at last to the old gas station on Woods Cove Road.

They were content to hold their services outside of town. They were used to the margins; heaven, they and their kind had spent their lives on them, living out of trailers on disability or scraping occasional work in the cornfields or for local concerns like the Church Pew Factory out at Rainsville.

As for the trouble, well, it just seemed to come with snakehandling. And anybody who regularly handled diamond-patterned Death in the Lord's name could take any number of the eggs and insults that were hurled during the services. In their grandparents' time, stirred-up bees had been released by disbelievers in the churches. Sometimes, churches had been torched. Brief infernos would light up the night sky, ashy smuts dancing on the breeze to put down gently on distant fields or shingle roofs.

Three nights a week, and for hours at a time, Glenn's growing congregation just sang over the car horns sounding down Woods Cove Road, or ignored the sneering faces glimpsed at the windows, joyous to be worshipping the Lord in a place they could finally call their own. As for Glenn, he had a purpose and a congregation, a congregation that gave him status—and regular funds. For a while, those closest to the preacher dared even believe the violence and the drinking was behind him for good and all.

Their forebears, mostly from the Scottish Borders and from Northern Ireland, had first come to the Southern Appalachians, the frontier uplands of West Virginia and the Carolinas, Southern Kentucky and Eastern Tennessee, the hills of Northern Georgia and Alabama, as pioneering settlers in the 1700s. They had brought with them little more than the clothes they stood in, some livestock, a few tools and the King James Bible.

Their isolated lives were parcelled out in work: clearing the land, building homesteads and churches with tramped earth floors, raising their stock for milk and food, hides and lard, and growing potatoes and corn. The rest was sleep and prayer. They kept to themselves, and called themselves Holiness people after their strict beliefs. They kept Christian houses, read from the Scriptures, abjured alcohol and tobacco, lies and "backbiting."

They dressed with a sober modesty. In the same spirit, later generations favoured shirts buttoned to the neck or workman's bib overalls. Citing Corinthians, the women mistrusted jewellery and wore their hair long and unstyled except perhaps fastened in a modest bun. In the 1950s, they wore thrifty dresses made up from the cheap floral patterns printed on the sacks of canny feed manufacturers.

The twentieth century, however, increasingly set the Holiness people at odds with the American way. Unversed in commerce and contract, they had too often sold their upland plots, and their holy self-reliance into the bargain, to timber and farming interests for a pittance. The only work was elsewhere, in the cities of Chattanooga, Huntsville, Knoxville, Atlanta and Nashville, or in towns like Scottsboro and Fort Payne in Alabama, Cleveland in Tennessee and Rome in Georgia. They worked in the lumber mills and coal mines, in construction and on the railroad, in the stores, factories and restaurants. They drove trucks. And they felt defiled.

The modern world had touched, even tempted them, with its abominations. Society was fallen, awash with the filth of homosexuality and the provocative attires of Jezebel: strumpet make-up, earrings and bangles; beer, whiskey and cigarettes; the unpardonable utterances of the movie houses, of radio and television; chewing gum, soda pop and coffee; and, latterly, rock music, drugs and even homosexual marriages. For the Holiness people, the twentieth century was one long crisis. Seeing their kind being swept away on the foul currents of modern

American life spurred them to return with renewed spiritual vigour to their rickety churches, where living right became once more a simple thing enshrined in the unchanging Scriptures. It was here, in the first decade of the century, that men among them first found instruction in the final utterances of Mark to take up serpents. Snakes were evil; they embodied the century's very mood. Handling them meant perfect victory over Satan. Handling them was to rise above the wickedness of modern America.

But Glenn Summerford's redemption was to prove short-lived. By October 1991, the bad old ways that he had disavowed at Kingston nine years before had reclaimed Glenn with a vengeance. In the aftermath of some bad bites earlier that summer—notably from an exotic, a Mojave Desert rattlesnake out of Arizona that somebody had got off of a dealer—the preacher had been going under, leaving vodka bottles, crumpled Winston cigarette packets, trashy rented videos with dues to pay, serial infidelities and too many public brawls bubbling in his wake. For a few days, a bite from an eastern diamondback had turned his vision yellow. Too much booze and strychnine, too much suspicion, and too much plain rattlesnake venom in his blood had finally pushed him over the edge, pushed him to park the car out of sight one Friday evening so nobody might guess them in, to lock the front door and drag his wife to the shed where the snakes were kept, a gun to her head.

I took a motel room just off the highway outside of town. The room smelt of stale cigarettes and lingering pizza. I dug out the phone book and found, between Chiropractors and Cigar, Cigarette and Tobacco Dealers, entries for fifty-one local churches of a bewildering range of denominations. There was African Methodist Episcopal, the Assemblies of God, Baptist, Catholic, Church of Christ, Church of God, Church of Jesus Christ of Latter Day Saints, Episcopal, Interdenominational, Jehovah's Witnesses, Lutheran, Methodist, United Methodist, Nazarene, Pentecostal, Presbyterian and Seventh-day Adventist, but there was nothing under Holiness. Pastor Frank, Glenn Summerford and their like evidently inhabited a world outside the phone book.

The phone rang.

"What's happening?" a voice asked me. Male.

"Who do you want?" I asked, assuming a wrong number.

"You," he replied. I laughed. "After a good rubdown tonight?" he asked, cutting to the chase. I dropped the phone and went out in search of somewhere to eat.

Geno's Pizzeria, down by the lights on the edge of town, seemed safe enough from unsolicited rubdowns. Not that it was the kind of place to invite conversation. Pale blondes served large men who stared at television screens from under baseball hats advertising agricultural machinery. Occasionally, they wiped beer froth from their moustaches with the back of their hands. I turned to Scottsboro's *Daily Sentinel*. "Dear Abby," a correspondent wrote on the letters page. "My husband and I have a difference of opinion about what the date on a carton of milk represents. He says it is the 'sell by' date, and I say it indicates that the milk is good until that date. Would you please consult your experts and let us know the answer."

The man who walked through the door just at that moment—big boots, black beard, bandanna, khaki flak jacket—seemed less the sort to respect sell-by dates on cartons of milk than to grievously molest the dairy herds from which they came. Until he walked in, the other newspaper item that had caught my eye—Major Public Auction! Complete Liquidation of a Rocket Fuel Manufacturing Facility—had seemed most improbable. Now, the market was all too apparent. When he placed a bag on the counter, I expected him to growl *Fill it from the till, bitch. And don't try my itchy trigger finger.* I did not expect him to tell the waitress, perfectly respectfully, that she had given him Thousand Island dressing.

"And I actually asked for Italian, miss," he added. "Mind if I ask you to change it?"

So, I mused, driving back to my motel, Scottsboro folk were a bit fastidious about milk and burger dressings. I just couldn't square that with people who regularly got bitten handling snakes of their own free will and then refused hospital treatment to die in agony—an estimated one hundred of them since the first decade of the century when snake-handling had begun to be practiced, without apparent precedent but with a fervent enthusiasm, initially among the country churches around Chattanooga, Tennessee, about fifty miles to the north of here. As the spring gnats danced in the headlights, I imagined the most recently dead, Melissa Brown in Middlesboro, Southern Kentucky, just seven months earlier in August 1995, being welcomed by earlier generations of Appalachia's snakebite dead into an afterlife that they all believed in with the kind of untroubled, everyday conviction normally reserved for concepts such as a roundish earth.

Melissa Brown had been bitten just below the elbow at the little church out the back of Middlesboro by a four-foot-long timber rattler, one of some fifty snakes being handled during the morning service that Sunday. Fifteen minutes after the bite, she had lost the use of her legs. She had laid up at her preacher Jamie Coots's apartment in Bella Gardens until Tuesday morning, almost forty-eight hours later, when her husband, Punkin, unsuccessfully urged her to think of their five children, ranging between sixteen months and nine years, and allow herself to be taken to the hospital. "She suffered hard the whole time," said Jamie Coots, suffered until she lost consciousness on the same afternoon, when an ambulance was finally called at 3:38 P.M. By that time, however, Melissa Brown was in full arrest and was pronounced dead on arrival at Middlesboro Appalachian Regional Hospital.

Wherever Melissa went that Tuesday afternoon, she had no doubt what she'd find there: joy unspeakable in the arms of the Lord, and all the snakebite dead whose example she'd followed. People she knew from the church revivals and homecomings that attracted Holiness communities from all over the Southern Appalachians, days of prayer, song and hog roasts in churches and summertime brush arbors, simple shelters raised in fields and on hilltops. There was Bruce Hale who had died just six months before her after taking two bites from a rattlesnake at the New River Free Holiness Church outside of Lenox, South Georgia. There was Jimmy Ray Williams who died in Tennessee in 1991, and his father, Jimmy Ray senior, who had died in the same county as his son eighteen years earlier, but after drinking strychnine. There was the handsome preacher Charles Prince who'd grown up with rattlesnakes under his bed (Tennessee, 1985), Mack Wolford (West Virginia, 1983) and John Holbrook (West Virginia, 1982). There was Claude Amos (Kentucky, 1980) and the three who died in 1978.

Further back, there were those whom Melissa knew only by repute or through familiarity with their sons and daughters: happy Lee Valentine (Alabama, 1955) and, three weeks before him, snake-handling's founding father, George Went Hensley. Hensley, whose name sounded more like a direction, was said to gather up serpents in his arms "like a boy would gather stovewood" and claimed to be able to cure the dying and to walk the Tennessee River. He finally succumbed to an eastern diamondback near Altha, Northern Florida, after surviving some four hundred previous bites that had left him "speckled all over like a guinea hen." There were the four who died in just six weeks of 1946 within just a few miles of each other near Cleveland, Tennessee—Joe Jackson on July 13, Henry Skelton on August 17, Walter Henry on August 25 and Hobart Wilson on September 2. There were the three who had died in September 1945—Anna Kirk in Virginia, Lewis Ford in Tennessee and George Coker in Kentucky—and Johnny Hensley, Maudie Lankford and Jesse Coker in 1944. From August 1940, there were Martha Napier and Jim Cochran, who both died in Kentucky. Melissa was in the company of the faithful. She had come home.

⌣

"Lift the lid," said Glenn as they stood over the second box. "I said lift it." Darlene's left hand was now throbbing so bad from the bite that she did so, gingerly, with her right hand. A charge ran through the box's contents; the snakes drew themselves tight.

"Tell you what," said Glenn, kneading the barrel of the pistol in the nape of her neck. "You handle that snake right there," he pointed to the larger one, "without it bitting you, the Lord says I'll let you live." But the Lord, Darlene knew, had nothing to do with this. If the Lord was anywhere, he was certainly not here. She tasted bile. She was in danger of joining Appalachia's posthumous rattlesnake bite club.

Only thing was, she was not willing.

3. AUSTRALIA 1

If a snake bites you in this country, instant death follows. One of the most deadly and common snake's bite is so bad that the person bit only shivers and falls dead immediately.

—Letter from William Coke, a guard officer posted to Australia,
to his family in England in 1827

OCTOBER 1996

In the latter days of September 1801, an evident gentleman, but of a singularly protesting and cantankerous sort, was put aboard the *Atlas* at Cork. Sir Henry Browne Hayes was a notorious Irish prankster and dandy who had, in a recent and ill-advised development, become a convict. The *Atlas*, a transport ship, was headed for New South Wales with the first fair wind.

Earlier that month, when Browne Hayes's death sentence was commuted to transportation to the thirteen-year-old penal colony, history might have expected a chancer's sigh of relief from him. In fact, the spared man was mortified, and not because the usual convict terrors unduly concerned him. Hang forced labour, floggings, terrible victuals, home-sickness, Australian humour or even the awful parties they were heard to hold down there, he might have declared. In fact, the irrepressible Browne Hayes might have turned the whole episode into a picaresque jaunt as befitted his colourful life had it not been for reports that the place was crawling with the things he dreaded more than anything. Snakes.

By the time the *Atlas* left Cork for Rio and the Antipodes, Browne Hayes had already recommended himself—and his considerable private income—to the vessel's venal captain, James Brooks. Captain Brooks took time off from attending to the 2,000-gallon cargo of rum he would sell at vast profit in Sydney to arrange for Browne Hayes's relative comfort, and so line his pockets further. The other prisoners fared less well. In the course of her nine-month voyage, the *Atlas* would lose sixty-five convicts, over a third of her complement, to disease and deprivation. Prisoners spent much of the crossing shackled below in a mephitic fug of disease and excrement. An enraged reformist wrote of the convicts'

> bad and filthy bedding; some not having half the covering of their bodies; the privation of the nutritious part of their diet, by scumming the fat off their Brooths; the defect of their Cloathing in the most intense cold...complete and painful Testicular Ruptures hanging towards their knees...the asthmatic and swell'd or ulcerated legg'd subjects . . .

A comfortable cabin and the best available food allowed Browne Hayes to look beyond the rigours of the great sea voyage, but the lack of distraction was perhaps not in his best interests. A Testicular Rupture or ulcerated legg might have been just the thing to keep his mind from lingering on what lay in store for him.

Lurid reports of the snakes to be found in Australia had been reaching Europe from the very first days of Australian settlement, and would colour missives home long into the nineteenth century. When Alexander Marjoribanks, a gentleman traveller, arrived in Sydney in 1840, he did as tourists should and visited nearby Botany Bay, where the first convict fleet had originally come ashore in 1788. He wrote,

> On coming to the bay I looked in vain for the great variety of flowers from which it obtained its name. In fact I saw more serpents than flowers, so that I should have felt inclined

to have called it Serpentine Bay. Talking of serpents, I may mention that the bite of most of them in that country is almost instantaneously fatal.

In 1861, Horace Wheelwright wrote in his *Bush Wanderings of a Naturalist*:

> What the bushman has most to dread in the Australian bush are the snakes. I do not believe any part of the world can be more infested with these reptiles in the summer season. Let him walk where he will—in the depths of the forest, in the thick heather, on the open swamps and plains, by the edges of creeks or water-holes...nowhere is he safe...at any moment he is liable to tread upon a deadly snake, coiled up in his path...watching him with his basilisk eye, ready in a moment to make the fatal spring.

In 1910, the Reverend John Flynn, who would go on to found Australia's Flying Doctors, published a pamphlet called *The Bushman's Companion: a Handful of Hints for Outbackers*. The extensive section "What To Do In Case of Snakebite" suggests that anybody in a position to give first aid should "stab in with knife all round bite," "suck for dear life," "rub in permanganate of potash," "don't allow patient to sleep," and administer "black coffee—very strong." But since his pamphlet devotes the greater part of its length to the inclusion of extensive hymns, "Auld Lang Syne," lessons from the Scriptures, directions for making a will and, finally, the Order of Service for the Burial of the Dead, Flynn does not appear to have had much confidence in the medical treatment his publication counselled.

As the *Atlas* bucked across the Indian Ocean after putting into Cape Town, Browne Hayes also had ample time to reflect on the incautious actions that had brought him to this terrible pass—"the queer matrimonial speculation," in the coy phrase of the author of a paper addressed many years later to the Cork

Historical and Archaeological Society, "which qualified Browne
Hayes for free passage to New South Wales in 1801 and for per-
petual residence there."

Sir Henry, it seems, had quite fallen for one Miss Mary Pike,
a young Quakeress, when visiting her father at home in Cork
one evening in July 1797. It is rumoured that Mary was both
beautiful and spirited, and she was indubitably worth £20,000.
The combined effect of personality, looks and wealth which had
long captivated the young bucks of Cork society, all but undid
Sir Henry and inspired him to recklessness. He resolved to attempt Miss Pike's abduction.

Late on the night of July 22, Mary was roused from her bed
by a servant with urgent news: word had been received that
Mary's mother had fallen ill at the house of a relative, and
wished to see her daughter. A carriage was hastily made ready
and Mary set off, only to be stopped on the Glanmire Road by
five masked and armed men. As Mary's coachman fled into the
night to raise the alarm at a nearby hostelry, the gang removed
her to another carriage whose driver was none other than the
scheming Sir Henry.

The knight hastened to his own house, where a priest was on
hand to perform the service of marriage. Mary resisted. With a
gallantry of sorts, the heartbroken Sir Henry proposed to shoot
himself if she continued to do so. Which she did; and he didn't.

The ensuing stand-off continued throughout the night, not
least because the authorities never arrived to terminate proceedings which had long since descended into farce. The magistrate's men had checked the Browne Hayes driveway in the
early hours—strange goings-on had been reported locally—but
soon moved on, having found no fresh horseshoe marks leading, tell-tale, towards the house. Sir Henry, it was to emerge,
had had the horses shoed backwards.

By the next day, Sir Henry had released Miss Pike unharmed.
It had rapidly become apparent to him that she was quite un-
charmed by his reputation as an amateur actor and by his

amusing stories of the famous silk canopy, his dandy's signature, which he fitted to his tent when on exercise with the Cork county militia. If this surprised him just as much as her lack of enthusiasm for the abduction had—he thought she might at least have found the enterprise exciting—he was enough of a realist to accept that his attempt to win her hand had self-evidently failed. So he cut his losses and made good his escape.

From the descriptions that accompanied the 500 guinea reward offered shortly afterwards by Mary's family for the abductor's capture—"rather fresh-coloured, a little pock-marked, and brown hair, with remarkable whiskers"—it is not hard to see why this well-connected, young Irish beauty turned Browne Hayes down. Still, she must have admired his chutzpah three years later when, tiring of the chase, Browne Hayes finally gave himself up to an old family retainer, a Mr. Coghlan, preferring to see the reward stay within the Browne Hayes household (although some have uncharitably suggested that the payment to Coghlan was less a gift than in settlement of an outstanding debt).

Sir Henry began his life in New South Wales in a Sydney jail. But on his release, he was given "special convict" status, and "suffered to be at large without the formality of a ticket-of-leave." In August 1803, he paid £100 for a hundred-acre plot on the southern shores of Port Jackson, latterly Sydney Harbour, a partially farmed spot above a horseshoe bay. A sandy beach looked out on to black rocks where the spray broke, with views across to the north shore. He called his new home Vaucluse.

It was an undeniably handsome setting, but Ireland it was not. In a frenzied attempt to create an impression of home among the alien tangles of Australian bush, Sir Henry sent extravagant shopping lists back to friends and supporters in Ireland. One, headed "flower and seed list," which started with the likes of "eight pottels early beans," "half pound prickly spinnage," "two pottels furze seed for hedges," "salsify" and "sperritt," ended with a poignant request for "snails" and "hedgehogs." And this from a man who had survived a death sentence and countless brawls, and would go on to make ship-wrecks a personal speciality.

Sir Henry's fellow convicts, however, soon proved themselves no respecters of his property, howsoever he adorned it. Aggrieved ads appeared in the *Sydney Gazette,* bearing his name: "REWARD OF TEN GUINEAS: WHEREAS several highly ornamental trees of Honeysuccle and She-Oak have been lately cut down on the lands of VAUCLUSE...supposed for boat building, as the crooked parts were only removed..."; and "TEN GUINEAS RE-WARD" for information about the person "who, on Tuesday or Wednesday night last robbed my garden, at Vaucluse, of a quantity of Water-Mellons; and on Thursday being fired at, left Bags and Bludgeons behind."

There were, however, far worse trespassers than thieving convicts to contend with at Vaucluse. "The part of the country where I live," he declared despairingly to an acquaintance, an army major, one evening in 1804, "literally swarms with ven-omous serpents; there are black snakes, brown snakes, gray snakes, yellow snakes, diamond snakes, carpet snakes..."

The newspapers were hardly reassuring. In 1804, the *Gazette* carried the typical story of a "fine boy" who had been bitten in the left arm by a large black snake while tending his father's stock on the Hawkesbury River to the north of Sydney. "Grow-ing sick and faint soon after, the poor little fellow went home, to chill with horror the hearts of his afflicted parents, who had to witness his almost immediate dissolution."

Sir Henry might have appreciated the kind of advice that I would come across in an Internet discussion group almost two hundred years later. On my computer screen, somebody called Juanita explained that she was seeing four or five snakes a day in her Seattle garden, sightings she didn't in the least enjoy. She was variously advised to "get a giant purple stuffed cartoon snake and leave it around until there is no reaction to seeing it or handling it," and to "put a snake picture on the refrigerator door." "Give your snakes a nice name," said another sympathetic adviser, "such as Fred or Suzy. When you see one, say hello to it before backing away." The suggestions continued: "hang out at your local petshop"; "find a snake owner who is willing to let you meet one"; "go to the zoo and watch all the different species"; and finally, "you would do well to consult a good psychologist." Without therapy, this item of advice went on, Juanita soon wouldn't be able to "walk past shrubbery that might contain the dreaded object," "read an uninspected newspaper or magazine," "stand being near anything remotely resembling the dreaded object (such as earthworm or rope)," and, finally, "hold down a job or have a social life."

Juanita's snakes, of course, were not venomous. Nor did they take to turning up indoors, as Sir Henry's did soon enough. He continued,

> Now, so long as they confined themselves to the lawn and garden I did not mind. It was bad enough to have them there, but, with caution, I could avoid them. But the brutes have lately taken to invade the house. We have killed them in the verandah, and in every room, including the kitchen. Last night, I found one, six feet long, and as black as coal, coiled up on my white counterpane, and another of the same dimensions under the bed.

Sir Henry admitted to his friend that he had been suffering agonies of mind for eleven days. In the absence of good psychologists, he sought celestial help instead. "I have been praying," he explained, "from early dawn to dusky night, almost without intermission, to my favourite saint, St. Patrick. But he seems to take no more notice of me, nor of my prayers, than if I were some wretched thief in a road gang."

Later that year, St. Patrick finally seemed to relent. At least, a solution recommended itself to Sir Henry that, the knight felt sure, was divinely inspired. A series of letters were promptly dispatched to Ireland. In one, an agent was engaged to deliver 500 barrels of best Irish bog to the docks at Cork. In another, addressed to all ship's captains putting into Cork en route to New South Wales, it was explained that the bog barrels were priority consignments required by the colonial authorities for urgent botanical experiments. The captains were instructed to transport the barrels without delay; space might best be made for them, the letter suggested, by removing an equivalent weight of ballast from the ships' bilges. Back at Vaucluse, a ditch measuring six foot wide and two foot deep was dug in readiness by hired convicts around Sir Henry's home.

We build moats in our minds to keep snakes from us. Sir Henry, reckoning that a soil banished of snakes by the heavenly St. Patrick might at least put off Australian ones, was going one better and building one around the periphery of his property. He was following an ancient tradition, albeit on a characteristically flamboyant scale; in medieval Europe, it was believed adders could not enter chalk circles. In Ancient Egypt and India, snake-shaped bracelets were worn as protective amulets. Until recently, hessian ropes were regularly slung around the foot of African bush tents in the belief that snakes would not cross them. And in the American West, cowboys made such magic circles by throwing their lariats around themselves as they bedded down on the range. Gruffly, these cowboys claimed that the coarse horsehair from which the rope was made acted as a barrier against approaching snakes. That this was patently untrue most cowboys privately accepted; snakes routinely rub up against rough surfaces. But these were hard, practical men who, I suspect, were not minded to admit what the lariats truly became to them at night-time; a kind of amulet, a superstition, a reliance on older faiths.

The world has long been spotted with such magic circles. Some of them, like the faint imprint in the dust of a cowboy's rope, only lasted until the wind got up or were even erased by the vibration of departing hooves. Others, snake-shaped bracelet charms hammered from metal in earlier millennia, look

destined to outlast man. When new foundations were dug to
support a sagging verandah at Vaucluse House, Sydney, in 1928,
workmen uncovered a seam of chocolate-coloured earth that
arced through the light, sandy loam on either side; it was
Browne Hayes's magic circle, created over a century before by a
class of convict less fortunate than Sir Henry. As Sir Henry had
instructed them, the convicts had gathered at Vaucluse House
on the morning of St. Patrick's Day. They set to work, shovel-
ling the turf, recently landed from Ireland, into the ditch. A
party of gentlefolk who had arrived by boat to lend their support to the proceedings looked on as Sir Henry broke out in
snatches of song in praise of St. Patrick. In the afternoon, when
the ditch was filled, a convict who shared Sir Henry's fear of
snakes begged that he might collect the remaining scraps of
magical turf for his own use. "Take it," Sir Henry replied, "but I
would rather have given you its weight in gold." Later that day,
Sir Henry and friends celebrated his new security from snakes
by dining outside in the shade of a sail loaned for the occasion
by a brig lying in the harbour.

Snakes, it should be said, are occasionally seen around
Vaucluse. The magic may have faded over the years, but it is
the pattern of circles over the years, large and small, temporary
and permanent, to which Browne Hayes's ditch singularly con-
tributes, that testifies to the nature of man's attitude towards
snakes. We are minded to see them not as flesh and blood, like
other creatures, but as half-lit spirits, visitors from another
world with a disturbing agenda for humankind.

Closed doors, we have somehow always known, will not keep
snakes out. What we use to deter other menaces—traps and
padlocks, walls and weapons, poisoned baits—are acknowledged as having negligible effect. In the case of snakes, the
usual physical restraints seem inappropriate, even one-
dimensional or simple-minded. So we rely upon talismans and
fetishes instead. And magic circles are just one of the charms
with which we have stocked our defensive armoury.

In the ancient world, it was long the tradition to repel snakes by the use of garlic. Even in the early twentieth century, it was the practice at nightfall to smear the ropes of tents in the African bush with it, as if to keep vampires at bay. Vampires, which visit humankind as bats, might as convincingly, I have often thought, come as snakes; certainly, these two denizens of the half-world have more than garlic and fangs in common. Vampires were believed to have similar trouble entering circles. Like vampires, who fall back in the face of the crucifix, snakes have long been regarded as the devil incarnate. Only a stake through the heart, the lore goes, could kill a vampire; in rural England, vipers were traditionally dispatched by being nailed to a tree until their death at dusk. The Lamia, the demon female of Greco-Roman mythology which is generally acknowledged as inspiring the Gothic vampire, actually takes the partial form of a snake, "a gordian shape of dazzling hue" in the words of Keats's eponymous poem. Which leaves us with the words vampire and viper, their similarities resonating stubbornly in the mind for all their different etymologies.

I was going to Australia, too, and I was grateful for Browne Hayes's example. The extremity of his predicament was a comparative comfort to me. His fear lent my own a perspective—he provided me with a kind of emotional ballast. I realized I'd been making his type my speciality recently. Like drunk John Girling, like Darlene, like Juanita, he was just one of many figures I had now collected, contemporary and historical, to have suffered at the hands of snakes. I now recognized those hours in libraries, in archives and on the phone for what they really were; less research than a kind of recruitment drive to provide myself with comforting company for the journeys ahead. I had been signing up snake sufferers, press-ganging them into service as travelling companions who would lend me succour in the months to come by reminding me how much worse things *could* be. Their collective function, then, was to be eternally worse off—more

snake-scared, more snake-endangered, more snake-bitten—
than myself.

Browne Hayes had done me another favour; it was his story
which suggested I carry a charm of my own. Sure, 1996 was not
1801, Heathrow was not Cork and the Boeing 747 was not the
Atlas. The captain, moreover, would make pleasant in-flight announcements and clearly had no illicit rum interests that might
jeopardize the comfort of his passengers. But if things had
moved on, then the constant was the fear of snakes.

Clearly, Irish peat was impracticable. Anyway, I rather inclined to the old English reliance on ash wood; snakes, it was
believed, never went near it. To make me feel safe, I would take
a piece of ash with me. The afternoon before my flight, I took
my penknife and wandered out on to the hillside opposite my
home in Gloucestershire. A spinney of young ash trees ran
along the road at the foot of the hill. From a branch, I cut a
short, light length about the size of a bullet. It was in my
pocket, sap-green still, as I left for the airport.

4. AFRICA I

Almost a hundred years after the *Atlas* had transported Sir Henry Browne Hayes to Australia, another ship carried the rather more willing Henry Tarlton from Australia to Africa. And almost another hundred years later still, in the teacup-tinkling hotel lounge of a Scottish country town where he had recently settled after a lifetime in East Africa, Tarlton's grandson was recalling how his newly arrived grandfather had decided against waiting for the Nairobi train.

"Mombasa," Pete Bramwell told me, setting the scene. "Early 1898." He sucked on his pipe and scanned a laminated lunch menu through pebble glasses. Eyes that had once spotted elusive savannah game at a distance now strained for clarity among burgers in buns and toasted sandwiches.

"Waiting for the Nairobi train was not on," explained Pete. "It would have meant kicking his heels in Mombasa for months, maybe years, would waiting for the Nairobi train." The phrase had rather more epic connotations in 1898, when it meant nothing less than awaiting the local implementation of Stephenson's great invention. Line-laying Indian coolie labourers, hacking their way into the interior, were still two hundred miles short of the place they called Nairobi. At that time, the Protectorate's future capital was a motley collection of huts sheltering missionaries, pioneer farmers and railway surveyors.

Britain's great railway inland to Lake Victoria, at once intended to secure the Empire's shaky hold on Uganda, kill off the remnant slave trade and open up the interior of what would come to be known as Kenya, was mired down at Milepost 131, on the banks of the Tsavo River. Construction of stone bridge piers in the soft riverbed was exasperating the abilities of the engineers. Trains that supplied the railhead from the coast pulped great infestations of caterpillars against the tracks, creating a slick grease which deprived the wheels of friction. No amount of creosoting, meanwhile, seemed to deter the white ants from devouring the wooden sleepers. By March, earthworks were being washed away by the seasonal rains that push north from South Africa. Carts mired. Lions started eating the workforce.

This is how Pete Bramwell's Kenyan family history begins. Ninety-nine years before we meet, his maternal grandfather steps off a steamer from Australia to hear the stories hanging thick from the eaves of Mombasa's clubs, consulates and hotels; rebellion in Uganda, maneaters at Tsavo, the going price for the fertile farming lands north of Nairobi and delays on the railway.

"By all accounts," said Pete, vigorously cleaning out his pipe, "grandfather was not much good at waiting." Henry Tarlton counted his own grandparents among Australia's early European arrivals. He had the settler impulse in his blood, and was itching to leave his mark on the Empire's latest frontier, the vast, untouched interior of East Africa. One suspects, however, that he was relieved to hear of the railway delays. For one thing, it kept the competition at bay; he had not come all this way to engage in an unseemly scrabble for land. For another, what pioneer worth the name had ever arrived on the virgin frontier in this new-fangled train thing? Why not earn the land's respect, the land to which he had pledged himself, in the *getting there*, as settlers had always done?

So Henry Tarlton set about assembling a caravan for the trek inland. He would be one of the last to do so, defying progress as he ventured west towards the land he had promised himself. He would follow the rutted tracks of the slavers, surveyors and missionaries which, with the establishment of the railway, were to have disappeared under the rank, tangled incursions of young bush by the beginning of the new century.

Tarlton bought carts, teams of mules and bullocks, and hired herdsmen and porters. He supervised the stowing: cabin trunks, plate and furniture, ploughs and mattocks, old paraffin tins in which water was carried, his soap and shaving mirror, candles, matches and tobacco, tinned food and sacks of rice, soda siphons, boxes of seed, *pangas* and crowbars, crates of squawking chickens, the canvas fly tent, mosquito nets, mattresses and linen, the .303 and cartons of cartridges, and medical supplies including bandages, castor oil, iodine, Epsom salts, Stockholm tar for treating cuts and quinine for malaria. There were also a few mementoes and keepsakes from Australia, and the items which, so the advice went, the natives might accept as barter or even—in a tight corner—as sufficient appeasement; beads, cloth and the highly prized rolls of wire from which tribal bracelets were fashioned.

Tarlton left Mombasa in late May. The available maps were basic and the few features marked—"waterless desert," "pond," the occasional "warm spring" and "endless plain"—were discouraging. The weather, however, was reasonable. The clouds had lost their great, grey dewlaps and the rains were dwindling. The cauldron heat had dissipated a little but it would still, Tarlton hoped, bake dry the tracks they intended to follow. At Kilindini, the railway's base camp on Mombasa's outskirts, thirty-foot lengths of rail and sixty-foot viaduct girders were being landed from lighters. Coolies staggered ashore with boxes of bolts and fishplates for the track sections. As the caravan passed the verminous shacks and tents, the hospitals, workshops and warehouses, voices called out to them. Where were they going? And why? Wait a year, and they could do it in comfort. And though the *sahib* and his party might reach Nairobi before the railway, they might as likely not reach Nairobi at all. Hadn't they heard of bandits? Of the fevers—blackwater fever, sleeping sickness, malaria? The snakes? The Tsavo lions?

Inland, the same questions greeted them. They heard them from the European surveyors and supervisors, calculating gradients from the pooled shade of their parasols. They detected them in the curious looks of the work gangs as they cut a tortuous swathe through the baobabs and thorn trees or dug rich red cuttings into the hillsides; in the expressions of the turbaned coolies, preparing the final surface for the tracks and sleepers with an allotment care, and even in the manner of the sweatshiny men shouldering sections of track, hot as cattlebrands, who paused to watch them pass.

They were good questions, as the sickening sights and smells along the way told Tarlton: the fly-blown, bloated corpses of donkeys and bullocks that littered the wayside, victims of dehydration and sleeping sickness; the vultures that descended lazily upon the carrion, marking the railway's stumbling march into Africa; the stench of dysentery among the gangs; the malaria cases converging on the track-side medical tents, and the first signs of sleeping sickness among their own cattle.

They pushed on across the lensed horizons of the flat Taru Desert. The thorns of euphorbia and alone ripped at their clothes and at the forelocks of the pack animals. At the Tsavo River, a vast construction camp had sprawled along the bank. The camp smelt of sweat and spice, half-cured game pelts and cooking meat, kerosene and shit, but mostly it smelt of fear. In the evenings, the stench of it rose in rank, sickly waves from the tents that the workforce had surrounded with thick *bomas*, barricades of wait-a-bit, the supposedly impenetrable thorn bush, to reach the quivering nostrils of the two lions that had acquired a taste for humans, bringing them silently to their feet. In the course of the year, the lions at Tsavo would show a comprehensive disregard for wait-a-bit by slipping through it as if it were so much shredded gossamer, to take the lives, almost at will, of at least twenty-eight Indians and uncounted Africans.

Tarlton's party stocked up on stores, repaired the carts and rested the pack animals at Tsavo. They stayed long enough to hear the dusk grunt of lions around the camp, and the coolies fending off the fearful silence with their singing. From the survivors, they heard accounts of screams filleting the night and, in the morning, the furrow marks made by the victims' dragged heels that led to their grisly remains. How they had once found a man's skull and several of his fingers—on one of which he still wore a silver ring. And how, with characteristic colonial efficiency, ring and ring finger were returned, along with the rest of his remains, to his widow in India.

Nightwatchmen were posted to clatter old tins. Fires were stoked, illuminating the posters that offered a £100 reward to the maneaters' killers. "The Manager of the Uganda Railway," the posters read, "having been incommoded by the depredations of man-eating lions..." Officers on leave and hunters after reputations spent the nights drinking or stalking noises in the dark while the coolies, who could not afford guns or were discouraged for reasons of discipline from owning them, took to their tents or tied themselves into nearby trees for the night. Even the local tribespeople, who had fortified their bomas, primarily to protect themselves against the slavers who had been taking their people for hundreds of years, reinforced their defenses against this latest threat. To Tarlton, Kenya must have seemed less like the promised land than a shuttered-up world, paralyzed by fear where a malevolent nature ran amok.

By Tsavo, Tarlton's party had also come across the snakes, a slithering presence on the tracks or among the wheels of the carts when they made camp. They did not know the snakes' names, only that they tended to come in brick-red and olive-brown versions in these parts, and were regularly six foot in length. They also knew, if Tarlton's knowledge of Australian snakes was anything to go by, that they had good reason to fear them. To keep them out, they slung hemp ropes around the feet of their tents at night in the traditional manner.

Snakes were a regular sight at Tsavo. At night, the lanterns the workers carried served to warn them not only of lion shadows among the tents, but also of snakes at their feet. The chief engineer at the Tsavo camp was J. H. Patterson. Patterson, who would kill the maneating lions in the latter months of 1898 and so leap-frog history into legend, was also adept at dealing with snakes. When his manservant reported a snake in the sahib's tent one night—clearly, the hemp rope trick was not entirely dependable—Patterson "seized the shot-gun." He

> rushed to the tent, where, by the light of the lantern, I saw a great red snake, about seven feet long, gazing at me from the side of my camp bed. I instantly fired at him, cutting him clean in half with the shot; the tail part remained where it was, but the head half quickly wriggled off and disappeared in the gloom of the tent. The trail of blood, however, enabled us to track it, and we eventually found the snake, still full of fight, under the edge of the ground-sheet.

Once it was dispatched, Patterson's friend, Rawson,

> picked it up and brought it to the light. He then put his foot on the back of its head and with a stick forced open the jaws, when suddenly we saw two perfectly clear jets of poison spurt out from the fangs. An Indian *baboo* (clerk), who happened to be standing near, got the full benefit of this, and the poor man was so panic-stricken that in a second he had torn off every atom of his clothing. We were very much amused at this, as of course we knew that although the poison was exceedingly venomous, it could do no harm unless it penetrated a cut or open wound in the flesh.

Or—as Patterson's account clearly suggests he did not realize—unless it got in the eyes. Patterson was dealing with a red spitting cobra. Had the stricken snake's poison caught the clerk there, he would have experienced a pain that Pete Bramwell describes, from first-hand experience, like having the acid from a car battery tipped into the eyes.

It was the bite of such a snake that would kill Tsavo coolie

Amam Din in December 1898. For protection from the lions, he had been in the shrewd habit of squeezing into one of the portable steel water tanks to sleep. The tank entrances were so small as to frustrate the lions, who could only swipe at the opening with their great paws. One night, however, the unfortunate Amam Din's tank happened to be already occupied by a spitting cobra. The sheer terror that presumably prevented him from seeking medical attention meant that Amam Din had been dead for hours by morning. Rigor mortis had set in, and no respectful way to remove the body from the tank could be found. The water tank that was to have kept Amam Din alive, Patterson decided, would now become his coffin. He had the tank carried down to the Tsavo River where it was ceremonially floated off on the current. When the tank showed no tendency to sink, Patterson put four .303 bullets into it, and the body of Amam Din went bubbling to its rest.

It was an extraordinary finale, not least in the striking manner the Indian's makeshift obsequies resembled those he could have expected to receive in his homeland. While spiritual purification by cremation, ideally on the banks of the Ganges at India's holy city, Varanasi, is the preferred Hindu funerary rite in most circumstances, some Hindu dead are instead left to float downstream. These include babies, pregnant women, *sadhus* or wise men, those who have succumbed to smallpox and victims of snakebite. The snakebite dead are lashed, cruciform, to a rudimentary raft of banana branches and cast adrift on the holy river, which will wash away the poison and deliver them cleansed into the next life. From these strange similarities Amam Din might have taken solace, even if he did go to the bottom of an unholy river far from home, in a metal tank and riddled with posthumous bullets from his sahib's gun, to end up as feed for the crocodiles.

Tarlton's caravan moved on through a series of sunrises towards a horizon which swam among acacia trees. The men were glad for the sanctuary of the Scottish Mission at Kibwezi.

Doubtless, they marvelled at the views of the Kilimanjaro ice cap sparkling in the west. By early July, however, the last of the bullocks had succumbed to sleeping sickness. The mules shouldered the extra load. Despite widespread dysentery, everybody was made to walk. It was when they reached the settlement at Machakos, in mid-July, that they first began to believe they might make it. At the Athi river crossing, they were within two days' march of Nairobi. The exhausted caravan finally arrived at the collection of huts and canvas tents arranged in lines along the edge of a swamp where mosquitoes dangled in the evening air. There were a few stores, a telegraph office, a mule station and a land office. Here, they leased Henry Tarlton 20,000 acres out near Thika at 50 cents an acre. He called his farm Ndururumu, after the dam he planned to build there. Pete Bramwell's forebears had settled in Africa.

JUNE 1996

Dawn broke over somewhere. It could have been Sudan, Somalia or Kenya's Northwest Frontier. From the window of the airplane, a thrumming metal pod carrying us south through the high atmosphere, I could see the first pink fingers of light feathering the high ground, and the pencil-coloured shade crowding into the sanctuary of gullies and depressions. It was a landscape inscribed with the evidence of its own beginnings; the wet-season rivers and creeks sidewinding their watery escapes into the valleys, and the wind that had cut similarly serpentine folds among the granular, desiccated escarpments of the semi-desert. Thirty-nine thousand feet of altitude had obscured what little evidence of man there may have been upon this land, and the earth looked empty like it must have done to the first men whose ancient remains have been uncovered along the great Rift Valley, the black, shadowy fault I fancied I could see running across the far extent of my view.

From a perspective they had never had, high above the earth, I could see why the ancients had believed that it was the great serpent, the original being, who gave life to the world by carving out the river beds with his 7,000 coils. All over the world, they saw the serpent in the movement of river currents, in the coiling rain clouds, in the waves of the sea, and in the rush of wind through the tree tops or in the swaying dance of a sinuous tornado. When tremors shook the land, they imagined the serpent shifting its Atlas burden, the earth, across its coils. They acknowledged the power of its strike in flashes of lightning and, in the lunar eclipses, saw it devouring the moon like an egg.

Around me, passengers began stirring in their litters of headphones, glossy magazines and laptop computers, pushing their Zoro sleep masks on to their foreheads. Rogue clots of turbulence in the airstream littered our approach to Nairobi, nudging us in our seats as if with a collective fit of hiccups. Before landing, I brushed my teeth and, despite misgivings that it might take my mind places, the strange places its many users had complained of, I swallowed my Lariam anti-malarial pill as advised. Perhaps this was why, as the ground rose towards us, the brief tails of dust kicked up by a few wheeling zebras appeared purple. And the sunlit green thorn platforms of the acacia trees were bomas hovering over the brown plain like flying saucers from which yellow and green birds scattered at the roar of our passing. And the tarmac airstrip was man's brief moment in the untamed infinitudes of Africa. And it struck me that, if what they said was true, Lariam and black mambas might just prove an injudicious combination.

Thankfully, the dour process of entering Kenya—the lines, the routine questions, the shambling circulatory progress of the luggage carousels—had the effect of grounding my fitful imagination. I emerged from the airport building into a scrum of minibuses that were waiting to transport European arrivals to the national parks. Taxi-drivers and the more down-at-heel op-

erators circled, like jackals and vultures at a kill, awaiting the
scraps that might remain once the larger operators had gorged
their fill of pre-booked clients.

A taxi-driver approached me. "Amboseli Tour?" he asked.
"Three days Tsavo–Mombasa? Samburu, Big FIVE tour, no
problem."

"The snake park?" I asked.

"Snake park," intoned the taxi-driver without enthusiasm. He
had hoped to land a bigger job. "Why you want go there?"

"To see the snakes," I replied lamely.

"Buuuwwwrrruua," he said, as a shiver danced across his
shoulders. "NO TIME for snakes." But he agreed to take me.

He was a Giriama, he told me, and he had the aquiline,
Arabian features of those people of the Kenya coast. His into-
nation was strangely unpredictable but his eyes were friendly.
He had the endearing habit of turning his head full-on to address me while his taxi proceeded as if without him. Warning
him of impending collision was rapidly established as my re-
sponsibility.

We discussed the weather (it was the end of the rainy season), the football (England had gone out of the European Cup
to Germany in the course of my flight from London) and the local mangoes ("yes, YES, the best mangoes in the WORLD"). After this extended preamble, I asked him about snakes. A long
silence followed in which I could sense the Giriama's shame.
He was too busy urbanizing, acquiring strut and street smart-
ness to regard the beliefs of his village past as anything more
than a fanciful embarrassment. But I persisted.

"OK, OK," he laughed, turning to me and raising his hands
in surrender. Which would have been of no consequence on a
straight. I stretched out a hand over his shoulder to steer us
round the bend, and the Giriama gave me a courteous nod of
gratitude.

"When I was child, we believed snakes come for us," he said, the hooked index and forefingers of his free hand demonstrating snapping fangs as we sidewound through the morning traffic on Moi Avenue.

"Snakes could appear just so," the taxi-driver continued, throwing his clenched fist open in a demonstratively flat palm. "They could come out of the blue, bite you so you DON'T know, not till you're DEAD."

In less than a century, Henry Tarlton's swamp-side settlement, in which it was said you could walk into a bank and credit elephant tusks to your account, had become a graceless but bustling city of shanties and European suburbs, government offices, aid organizations and press bureaus, markets and tour agencies, curry houses and hotels, mosques and cathedrals, and the odd etiolated skyscraper, a civic trophy sprouting from the plain. On the pavements, fruit-sellers squatted by neat piles of mangoes. Young men hawked wood-carved game animals and the bead bracelets they had slung along the shop-rails of their arms. A supermarket promised "Happy Shopping with Coca-Cola." A billboard peddled the apparent joys of Haliborange Fish Oil: in the morning heat, the thought made me feel nauseous.

"Some people even thought snakes were their relatives," said the taxi-driver, warming to his theme, "back from the dead." Snakes were not only agents of vengeance, then; they were also, antithetically enough, family. I was reminded of a memorable episode in 1882 that writer and anthropologist W. C. Willoughby had recounted. While travelling near Mpwapwa in what is now Central Tanzania, a few hundred miles south of Nairobi, Willoughby was shown the grave of "Dr. Mullens, former Foreign Secretary of our London Missionary Society," by his Zanzibari porters. As they approached, they "saw a large snake sunning itself on the loose stones that had been heaped upon the grave to ward off hyenas, wild pigs and other burrowing beasts. 'There he is!' whispered the Zanzibaris in awestruck voices." They were referring to the doctor.

In many cultures, snakes have long been regarded as embodying the souls of departed ancestors. Once, when a certain Mr. Henry was travelling in America's Indian territories in the 1760s, he attempted to kill a rattlesnake that had strayed into a camp of the Ojibwa tribe. Alarmed, the Indians stayed his arm, called the rattlesnake "Grandfather," blew appeasing smoke over it, asked it to take care of their families and even wondered whether it might have the power to cause the local British commercial agent to resupply their canoe with rum.

One anthropological theory, ingeniously simple in explaining the ubiquity of this belief, is predicated on the most graphic manifestation of the putrification process. The maggots that tend to emerge from corpses in tropical climates, the theory goes, resemble baby snakes; at least, it is persuasive to suggest they may have done so to our ancestors. This, then, was how the souls of the departed moved on. To Willoughby's Zanzibari porters, it was plain that the departed Dr. Mullens had simply done the conventional thing and come back as a snake. And as befitted his status—he was, after all, the former Foreign Secretary of the London Missionary Society—he had come back as a large one.

It was even believed that distinguishing mutilations or markings that had occurred during the human phase were replicated in the reincarnated snake.

"If such a snake has a scar, or has lost one eye, or is lame... a late-lamented dweller in that *kraal* had a similar blemish, and this is his reincarnation," wrote the intrepid Reverend Henry Callaway, who studied African tribal beliefs in the field during the 1860s.

I might have asked the taxi-driver how one distinguished between one's well-meaning ancestors and those unpleasant snakes that were sent to get you, but Reverend Callaway had been extremely clear on this issue. "Divine snakes enter a house and remain quietly coiled," wrote the Reverend. They look after their families and apparently warn of danger by lying on their backs whereas ordinary snakes "enter a hut apprehensively, glancing from side to side…and may be killed at sight." These were useful clarifications; killing the likes of one's reincarnated grandmother by accident required a sustained sacrificial appeasement—two sheep at the very least, it was said.

"So," I pressed on, trying to square these apparent contradictions, "some snakes want to protect you while others intend you harm?" But the taxi-driver was tiring of me.

"So many questions," he admonished me gently, pulling up by the snake park. "Look. Snakes CAN be many things. Not just animals, not just spirits, NOT just relatives. Snakes have POWER. I don't like them."

I nodded, chastened, and handed over some banknotes. "Will you wait for me?" I asked.

He looked pained. "Not here," he said apologetically. "I don't like it here." I shrugged and he drove away to look for safer fares, his battered blue taxi bobbing down the stream of morning traffic. Despite adulthood and the secularizing effect of life in the city, my taxi-driver remained the child who believed in the power of snakes.

"No member may attempt to enter the snake pit," read the sign on the approach to the snake park, which struck me as an unnecessary injunction. The snake park was a lowly concrete building that might have housed municipal toilets if transported into the setting of an English provincial car park. The tail of a crocodile of blue-smocked children was disappearing through the entrance. Inside, the discipline evaporated. Whinnies of fear escaped the children as they took in the glass-fronted pens recessed into the walls along three arcaded sides, and the central feature, the sunken concrete snake pit, open to the sky, boasting a few shrubs and trees, and posted with the blunt caution "Trespassers will be poisoned." An attendant stood in the middle of the pit, picking up harmless sand boas that drew screams from the children.

"Who wants to hold one?" the attendant asked, and the children shrank back as if touched by flame.

In the pens were arrow-headed puff adders, whorled spirals of brown and yellow, heavy and somnolent; a gaboon viper, exquisitely patterned with rectangular patches of pinks and browns; black cobras weaving across the glass like dancing tongues of smoke; and a tiny vine snake wrapped vermiform around the mottled branches it so resembled. I thought of the words of Callaway and the taxi-driver, and wondered for a moment whether I was among modern Kenya's ancestors. In that none of these snakes were lying on their back, no clues were forthcoming from Callaway. Some of the snakes were quietly coiled, but not in a convincingly ancestral way. I remembered that Callaway had also made the point that the species of snake was a reflection of human status; tribal commoners tended to come back as grass snakes or sand boas. In a modern context, then, did I detect in that slothful puff adder, irritable, overweight but possessed as it was of an impressive if malign presence, the spirit of some bullying downtown bank manager since departed? Or that of a shy but beautiful country child in the vine snake? Or were these merely snakes? My mind was straying in the heat.

A pen in the far corner, I noticed, had attracted a large crowd of children. They seemed transfixed by the snakes which had grown active in the heat. Long and sinuous, they seemed to pour themselves along the pen's arrangement of branches, or shrugged effortlessly across the floor with the motion of wavelets running up a beach. Some were seven feet long. They were olive brown along the tops of their bodies. The pearly white of their fish-scale underbellies extended across the sides of their heads, and gave their jet-black eyes a hypnotic prominence. But it was the heads that held my eye; after the fluidity of the long body, the striking suggestion of angularity as the head tapered like a coffin, a delicate skeletal composition that made the bulbous jaws of the vipers and adders seem crass and thuggish by comparison.

"*Dendroaspis polylepis,*" read the sign on the pen. "Neurotoxic nerve poison. Known as Africa's deadliest snake." We were standing before the black mambas. Or reincarnated tribal chieftains, as the awed Reverend Callaway had claimed. The black mamba has also inspired directors of snake parks, collectors and herpetologists, men who mostly tend to enthuse about abnormalities in scale arrangements and abstruse aspects of snake husbandry, to flights of uncharacteristic hyperbole, as if its fabulous reputation has helped it weather the reductive raids of twentieth-century herpetology. Deep in the memoirs of Africa's snake men are sudden descriptive blossomings that label the black mamba as the "Attila of Snakes." "The diabolical ingenuity of Nature," wrote F. W. Fitzsimons, who ran the snake park in Durban, South Africa, earlier this century, is "brought to the highest peak of perfection in this snake." Fitzsimons, who described these snakes in terms normally reserved for maneating lions, wrote that "an old black mamba, which has become a manchaser and stock-killer...loses his fear of our race and gets aggressive...and is more to be feared than the very devil himself." South African naturalist Eugene Marais went further still. "I had seen the menace of a wounded lion before," he wrote, "but he could not be compared with a mamba." Marais also claimed that a ten-foot mamba could inject a teacupful of venom into its victim while Fitzsimons declared just two drops of its venom to be "a fatal dose for the strongest man."

Something about these snakes had also silenced these Nairobi children. His sense of fun, however, finally got the better of a boy in orange, mud-spattered socks, and he swiftly sank his hooked fingers into the arm of a friend. His playful gesture, delivered with a sudden swoop of the arm, set the victim shrieking and scattered his classmates while the teacher flapped his arms in a vain bid for order.

To rural peoples within its sub-Saharan range, a bite from the black mamba meant sudden death. In Botswana, some tribes knew it as the Two-Step snake, and not for reasons of dance. In parts of South Africa, it was known as the Shadow of Death. Fit young men had been known to die from its bite within minutes. But it was not death they feared; death they could accept. Being mauled by a lion, charged by a buffalo or taken by a crocodile was never a pleasant way to die, but its self-evident manner at least meant the autopsy was for sure and had been so ever since the dawn of man. When the body was shredded, broken or plain consumed, it didn't need asking how the victim had died. What they truly dreaded was the unknown. The mechanics of envenomization (when the snake bites, poison is injected into the victim through internal ducts in the fangs, or in some snakes is expressed into the wound by means of grooves on the fangs' exterior) may seem simple enough to the scientifically minded, but the sickening process associated with snakebite remains baffling to many Africans. Snakes' headline characteristic—the ability to inflict sudden death by a seemingly innocuous injury—has shaped man's primary dread of these creatures. Settlers to Kenya in Henry Tarlton's time often noted the local reverence in which the healing power of syringes was held. Perhaps this was predictable; to tribal Africans, hypodermics represented snakebite in reverse, a kind of intravenous, magical salvation. As if the *mzungu*, the white man, had answers to the mysteries, his own brand of *dawa*, or magic medicine.

"Do you ever see them around here?" I asked the park attendant as he passed the black mamba pen.

"Not in Nairobi," he replied. "We're too high here. The best places are Tsavo and Voi, and down on the coast. They like the heat. But they're impressive, eh? You wouldn't want to get bitten by one."

I wondered if he knew of any black mamba victims.

"Truly," he said, "not too many of them survive. But there was one bloke, a snake man down at Kilifi on the coast. Called Barnsley. Something like that. He took a dreadful black mamba bite. And survived."

February 1997

When I first found Pete he had selected a quiet corner of the Banchory hotel lounge, and was huddled over a pile of old photographs in a bolt of spring sunshine. On the phone, he had told me how to find him with an honesty that bordered on a plea for charity; elderly, grey-haired and thin, he'd said. The description led me straight to the man with a beakish face lined like a weather-scoured landscape and framed by thin, lank hair. He sat alone.

"These Scottish winters," he'd added on the phone. "I can't shake the cold out of my bones till October. Then round comes the next one."

Over the years, Pete Bramwell had endured too many bush accidents and too many snakebites for his own good. In the 1980s, he'd even been run down on a Mombasa zebra crossing by a Landrover belonging, of all things, to the social services department. Pete shook his head, baffled by the patterns of history. What had brought him to Britain in the late 1980s had also taken his grandfather, Henry Tarlton, to Africa from Australia, and in turn had taken that man's grandfather to Australia from Britain; the colonial experience and a family adventure, history in the round, finally winding up in a Banchory hotel.

Early in our correspondence, Pete would reveal a longing for the life that age and snakebite had wrested from him before geography finally put it out of reach. He left messages for me which he concluded with the word "over," as if he were talking over crackly bush radios rather than to a Home Counties answerphone. He was forever recognizing the places on TV wildlife programmes where he'd picnicked or camped; Naabi Hill, the Olduvai Gorge, the Ruvana River. Even in genteel Banchory, Kincardineshire, Pete was wearing a jacket so old and scruffy it seemed to hail from that distant time. He wore it, perhaps, in readiness for the day that his beloved African bush might surge all of a sudden up Banchory High Street, past John Menzies, the bank and the knitwear shop to push up against the panes of the town's Burnet Arms Hotel, where we were drinking coffee and smoking, to swallow him up once more.

Among the photographs—of grandfather Henry, founder member of the Gentleman Riders' Association of East Africa, posing in his jockey's kit; of family and friends caught in sepia poses under acacia trees—one of Pete stood out. He proffered it as if casting to retrieve the past in which he appeared: twenty years old, and perched triumphantly on a fallen elephant with splendid tusks, his first ninety-pounder, shot near Taveta on Kenya's border with what was then Tanganyika. He carries a rifle, wears long dirty shorts and a smile that claims the world as his, beneath hair that is a mass of blond curls, Danny Kaye style. It was the late 1940s, when Pete's life was about breaking wild horses for buyers in the Belgian Congo, swimming in Lake Chala (after throwing in dynamite to clear the area of crocs), and collecting cobras and puff adders with his Uncle Alan, soon to be made head of the Kenya Medical Corps (Snake Section).

He was born in 1929 to Henry Tarlton's only daughter, Olive, on a wheat farm in Kenya's Laikipia Highlands, just a few months before an earthquake struck the family quarters in the middle of the night. His father, family lore had it, had plucked him from the cot as it slid into the yawning fault that divided the room. In his childhood, Pete saw plenty of snakes. During school holidays, he used to accompany his uncle on trips up-country to Embu to collect them for antivenom production. But it was not until 1945 that he first came close to a bite.

On New Year's Eve, the Bramwells had gathered at Malawa, the farm of family friends situated on a *tilapia* fish and trout stream that runs into Lake Naivasha. In the hours leading up to midnight, when old Granny Ullyat would fire off her old .303 at a minute before midnight, on the hour and a minute past ("to shoot out the old year," Pete would tell me, "and shoot in the new one"), family, friends and military convalescents billeted at the farm were playing "kick the tin." This game of tag led Uncle Alan into the moonlit paddock, his nephew in hot pursuit. Pete followed behind a *leleschwa* bush where, much to his surprise, he ran into an almighty shove that sent him sprawling. Stung by his uncle's sudden change of mood, he scrambled to his feet to see a large puff adder had nailed itself to Uncle Alan's forearm, its fangs sunk firm into the flesh. The shove had been well intended. Uncle Alan ripped the snake free and dashed its head against a rock. By the time they reached the house, word of the accident had spread and the dishevelled house party formed a hushed circle around Alan.

They lay him on the sofa. Fearing an allergic reaction, he refused antivenom, the antivenom produced from the poison of the very snakes he had collected over the years. He called for a knife with which to gash the wound. His teeth chattered as he lay open the wound with a long slash of the blade. Pete sat by him. Somewhere in the house, he could hear music. Later, he would remember how inappropriate the music had seemed at the time; Pete was fiercely protective of Uncle Alan. The colour had gone from his uncle's face. The arm had swollen and had a hard, hollow look, so much so that one was tempted to hit it like a drum. A thin trickle of discharge, mostly colourless but occasionally containing brown swirls of what Pete took to be blood, seeped from the wound, and the sweat that beaded on Uncle Alan's forehead was soon making the pillows damp.

His uncle was suffering hard—as Pete was to find out twenty years later, the puff adder delivers an excruciating bite. The swelling can be such that the only way to relieve the pressure is by carving the flesh to the bone. Uncle Alan lay unconscious. He stirred when the .303 went off on the verandah. On the

third explosion, he actually smiled. It was then Pete thought his
uncle might just be all right.

"He pulled through," said Pete, mining out the bowl of his
pipe with a match. "He married Maria. She was a South African
girl who had ridden the gold-bullion trains in Kimberley with
her father. They were employed as shotgun guards against bandits. Later, she became an excellent flamenco dancer..."

JUNE 1996

I had a ticket on that night's train to Mombasa. I had no hotel
room, so I spent the afternoon in Nairobi's Macmillan Library,
a place of intense study, enduring colonial echoes including
Queen Elizabeth's portrait and a moth-eaten lion-skin floor rug,
and comfortable chairs.

Between dozes, I read the notices; "BORROWED BOOKS," read
one, "must be protected from rain, babies and other hazards."
"INFECTIOUS DISEASES," read another: "Books must not be borrowed, or if borrowed, not returned, until the risk of infection
has passed." I scanned the papers. Kenya's most wanted criminal had been gunned down in Nakuru, a town in the north-
west. There had been fighting between ethnic Somalis and
Samburu herdsmen to the northeast, and a light aircraft flown
by a lone 72-year-old man had never reached home, and was assumed to have gone down in the thick bamboo forests of the
Aberdares. There was, I discovered, a Kenyan football team
called Black Mamba, and I happened on references to the fact
that the local bicycle was also known as the Black Mamba, perhaps because of the snake's fabled turn of pace.

At the station I found my name among Oakleys and Fields,
Kellys and De Jongs on the large varnished wooden noticeboard
headed Coach Seating and Berthing Allotment. Chastened by
my dinner voucher—"To avoid crowding the corridors, passengers should remain in their compartments until notified verbally

or by the chime gong"—I stared obediently from my cabin window as the train shambled out of Nairobi. The light was softening into a dusky wash as we set out across the Athi Plains.

The chime gong came, and we consumed our culinary colonial echoes—cream of asparagus soup, boiled rumpsteak and onions, and pineapple sponge pudding with custard sauce—in shifts. Later, lying on my bunk, it seemed I could hear every steel sleeper slipping singly beneath the wheels in the rhythm of the tracks. This was suggestive of so much labour that it exhausted me, and I drifted into a dream-littered sleep, a weird state in which snakes kept appearing, sleekly malevolent things which also happened to speak, reassuring me that they were in fact various great aunts, a great grandfather I had never met and a sibling I seemed not to know about, and that they wished me no harm.

I awoke with a lurch at Kima. It was here, I remembered, that a railway inspector called Ryall had been dragged from a siding carriage by a lion in 1901. I reassured myself that my cabin sharer, the shy Indian boy on the lower bunk, was more accessible, and drifted back to sleep. In the early hours, when the sound of the wheels hollowed and rose an octave to wake me on the Tsavo bridge, I thought briefly of Patterson and his coolie thousands, and of a bullet-ridden galvanized water tank.

At the time, I knew nothing of Henry Tarlton, who had followed hard on Patterson's heels and had then moved ahead to reach Nairobi years before that man's railway. Nor did I know that his grandson had taken this same train down to Mombasa one night in December 1963, ten days before Independence, with a plan in mind. Pete Bramwell was going to open a snake park.

5. AMERICA II

The guests heard the receptionist coming. Banging doors, urgent footfalls and a voice that inflected like a police siren marked her progress through the motel. When she was still far off, she distracted us from our muffins. As she burst into the breakfast area, sounds were tumbling from her puckered, glossy mouth and red patches were working their way across her cheeks. Nothing so thrilling, it seemed, had happened to the receptionist for weeks.

"We got weather coming our way," she sang out.

So motel life, Scottsboro style, was apt to pall. After a few days here, I could see that, but not so as even a change in the weather could bring the staff out in sub-sexual hives.

"We're in for a bit of rain then," I said.

"Ryan?" she squawked in a falsetto. I thought she was ignoring me for someone else, but when no Ryan answered, I cottoned on to what she had said even as she began repeating herself. "Rain? They got tornado warnings out for Jackson County."

"*And* Marshall and Madison," said a new arrival to the breakfast area.

"*And* Dekalb."

It was then that I looked up—to see the sky had turned the colour of the coffee I was pouring. You could have held the glass jug up to it and lost it against the swirling, preternaturally backlit murk that passed for a landscape. Trees and cars had become mere suggestions, partially developed images in the half-light. Passing motorists, who knew to heed tornado warnings, were clearing the highway, clogging the exits in their haste. They left their trucks, pick-ups and sedans on the motel apron, patted the hoods in passing in case they should not see them again, and burst into the foyer. They wore hunched, wary shoulders, as if the prospect of danger had turned morning errands, 10:00 A.M. appointments into death-defying dawn runs that only heroes would attempt.

"So you don't drive in weather like this?" I asked a short man whose pick-up had just exited the highway on protesting tires.

"No sir," he replied, wiping imaginary sweat from his brow and giving me a *where've you been?* look. "Not unless you want free flying lessons."

He had a gah! gah! for a laugh that was especially irritating at this time of the morning.

"You're in Tornado Alley here, particularly at this time of year. You got the cold air off the mountains coming down the Tennessee Valley and the warm stuff from the Gulf of Mexico funnelling up it. When they meet," he drained the Styrofoam cup in his hand, "things can get kinda hairy. Gah! Gah!"

As he crushed the cup, it went off with a submissive pop. Then the thunder hit, shaking the building with the sound of a heavy corpse being dragged unceremoniously down a flight of stairs. Fingers of lightning varicosed the sky.

"You hear me," the short man advised. "If a tornado hits, keep away from the windows."

No loose drops announced the rain. It arrived instantaneously, a stinging dance of cascading water that hammered against the deserted highway, and caught the baker in the midst of his delivery to the motel. His baseball cap went off like a gun, cannoning high into the jetstream, and his lank hair unfurled and streamed out after it, like a rock star's in a wind machine. Then a thunder clap in cahoots with a gust of wind mugged him, and his tray of loaves tumbled on to the forecourt, where they settled upon the pond that had already formed there. The high-sided loaves caught the wind and reached across the water like a fleet scattering before the storm to run aground upon the ramp in front of the foyer.

"Wet sandwiches!" shouted the exultant receptionist.

"Whooah! Give us both barrels!" yelled a trucker, daring the heavens.

"Got me sixty chicks waiting at home," said another man, looking at his watch. "Might just get drownded if I don't get back soon."

"We all might just get drownded if it carries on like this," said the pick-up owner. "Got any playing cards, honey?" he asked the receptionist. "Might as well get drownded playing five-card stud as most any other activity. Gah! Gah!"

The thunder grumbled over the hills, and squally shoals splattered at the window. Between the weather could be heard the low trill of shuffling cards and radio weathermen, the leafing of newspapers and the drumming of idle feet. A line of wet loaves had been arranged to dry along the coffee table.

"That was a tornado?" I asked.

"Not so far, honey," the receptionist answered. "That was just a gust of wind." The pick-up owner looked up from his cards, winked at me and extended his arms as banking wing tips.

"And where might you be headed?" asked the receptionist.

Kingston, Georgia, I told her. If the weather ever let up I had a church service to get to.

～

By October 1991, Glenn was acting real mean. Forever accusing Darlene of running around with other preachers, church members, anybody who came to mind. He was drinking too much, getting into brawls and spending money he didn't have in places he shouldn't be. Getting back late from the Lord knows where, a pattern of bruises and dried spittle.

She should have seen trouble coming her way on the Tuesday night. That was when Glenn came over all solicitous. Hardly his manner, was solicitousness. Certainly not after all his bickering earlier that evening. She'd swallowed some sleeping pills—all she wanted was some quiet from the cussing. Which was when Glenn grabbed her, forced her on to the lounger and stuck his fingers down her throat so she threw up the four Sominex.

"You don't want to do that, honey," he told her. Later, she'd understand it had all been for the benefit of their young son, Marty. Well, not for his benefit so much as for his attention. This was Glenn building himself a defense of sorts in case he ever needed one. Proving Darlene was suicidal. Making Marty a witness to that. For Marty saw him make her vomit the pills, Glenn made sure of that. Marty heard his dad talking to her softer than normal, then heard him tell how his mother had just tried to commit suicide. And all Darlene had wanted—but it was useless explaining what she'd wanted.

On the Wednesday, he beat her. She could tell he was spoiling for trouble when he picked up the phone in the afternoon to Gene Sherbert, a local preacher. He was always going on about Gene.

"Gene," he said. "You been messing with my wife?" He listened intently.

"Gene, I reckon you have. All I'm asking for is the truth." He listened again, for a while this time. Then he nodded, put the phone in its cradle gently as a baby and turned to his wife. His voice was low.

"Gene just told me you been messing together."

But, Darlene knew, Gene just told him nothing. Gene hadn't even been on the line. That business of inventing phone calls was Glenn through and through. Wasn't even convincing. Still, that didn't stop him from grabbing her red hair and beating her bad. She'd been cheating on him, hadn't she.

It was the Friday, however, that things got truly out of control down the far end of Barbee Lane. All morning, eruptions of rage had been raising the doves from the nearby trees. Even the dogs down the road at the Chambliss place pricked up their ears at the distant hollering. After one exchange too many, Glenn stormed from the room, swearing he would blow Darlene's brains out. Darlene knew the shotgun was stowed nearby, and was not for hanging around to find out whether her husband meant it. She took off up the hill at the back of the house. She didn't know much about shotguns so she kept running until she felt she must be out of range, out of range of a drunk, at least. She turned, panting, to see Marty stood between his parents, aiming his drawn bow and arrow at his father, her last line of defense. The sight seemed to shame Glenn. Either that or he realized a shotgun might not help the defense he had in mind. He lowered the weapon, and disappeared inside. Later, he even attempted an apology. There was a reconciliation of sorts. Even so, now did not seem like a good time for Marty to be around. Darlene asked her sister-in-law to have him for the weekend.

They dropped Marty off in the afternoon. They called on other family. Drove round a bit. Drank vodka and orange juice straight from the bottles as they did so. It was then Glenn told Darlene he was going to kill her. Threatening to kill her was no first—Glenn had menaced his wife with some terrible things over the years—but it was his manner that was unfamiliar. He was strangely calm about killing her, sharing his plans as if he was discussing improvements to the garden.

"There's an old hole I know in the side of the mountain out near Woodville," he told her. He would throw her in it, he said, and then he'd tell everybody how she'd run off with somebody. And he knew plenty of fellas she'd be quick enough to run off with. Wasn't just Gene Sherbert. The police would buy that.

"Better idea," he said suddenly as they followed the road towards Larkinsville. Evidently, he'd been pondering his options. He'd throw her off the river bridge instead, he told her. She'd disappear among the sloughs. Nobody would find her there except fat catfish. Strip her to the skeleton when she was rotten. He looked towards the trees cloaked in coppery leaves that slid by at the roadside, and nodded. "Yup," he said. "Better idea, that." Another first, never before had his previous threats run to reflections on how he might best cover his tracks after her murder. There had been times he'd wanted to kill her; now he was thinking about getting away with killing her.

Darlene wasn't up to being afraid. The vodka had settled upon her. She surrendered to its weight. She felt too tired for urgency. She felt mussed up enough as things were. She couldn't truly sense the harm he was planning for her, just the meanness of his words. They still hurt her, even after all these years. Silent tears slid singly along her eyelids to drop down her cheeks. She looked across at Glenn. Great, grey crescent moons had risen beneath each bloodshot eye. He stared out into the dusk from behind the steering wheel, the headlights of oncoming cars momentarily narrowing the eyes of his illuminated face, a face you couldn't forget. Jealousy and care, drink and poison had scored and mottled it. The teeth were in a state, but it still looked like a somebody's face.

Glenn picked up some burgers and took the road back towards Barbee Lane. As they approached the house, it struck Darlene that this wasn't the way to the river bridge. So she wasn't for the high jump, not this evening anyway. It was then that Glenn turned off into the woods. The wheels crunched over rotting, fallen boughs. A fall mist was rising from the river.

"So nobody will think we're at home," he told her, pocketing the pistol that had lain between the front seats. He wished for his wife to acknowledge his plan's clever touches. What he seemed to forget was that she was also his victim. But Darlene was strangely incapable of alarm. As they walked through the first fall of leaves towards the house, she even fell into step with him.

"I don't want no bothering this evening," he confided in her, padlocking the front door from the outside for the first time she could remember, and leading the way inside round the back.

Darlene followed like she always had. For this was where she lived.

They ate the burgers in silence. He began the moment they had both finished. But it wasn't the typical abuse he gave her; abuse Darlene could have taken, she was used to it. Instead, Glenn seemed to be offering an explanation. He was almost approaching apology. Here he was, trying to justify his actions and Glenn Summerford's actions didn't normally need much justifying.

"I don't want to live with you no more, but I don't want nobody else to have you," he explained, wiping his mouth with the back of his hand. "You have to die, you see, because I can't get remarried if we got a divorce. Because I'm a preacher and that would be adultery for me."

So that was it. Glenn wanted out. He probably had a girl waiting. But he was married to Darlene. And he was a preacher, remember, and preachers just didn't go round divorcing and remarrying. He'd had enough of her, sure he had, and not enough of the new girl, so the only way of ridding himself of the one and having the other was by killing Darlene. The more basic truth—that preachers weren't supposed to kill their wives—apparently hadn't occurred to him.

Glenn got to his feet, belched burger and picked up the pistol. With his free hand, he suddenly grabbed Darlene's hair and dragged her outside. The first few stars were appearing in the sky. Guessing he might be about to use the gun on her, she only had time to close her eyes. When she opened them, Glenn was fiddling at the door of the shed. As it opened to the musty scent of its contents, Darlene realized that he was going to use the snakes on her.

The snakes had occurred to Glenn out of the blue. River bridges and mountain holes were fine, but where the cops found bodies these days was pretty incredible. No place was safe like they used to be. As for the snakes, he somehow trusted them to do his bidding. Felt a bond with them. They'd brought him back to the Lord, hadn't they? Besides, Darlene had handled snakes for years, everybody knew that. The part of his brain that was bent on concocting a believable story recognized that getting herself bit was the sort of durn thing Darlene might just do. And snakes was how they used to do it, wasn't it; even Glenn knew about Cleopatra, the most famous suicide of all. And, finally, Darlene had been drinking plenty and was in no fit state to handle snakes just now. Darlene was going to get bitten tonight.

As he pulled her into the shed and stood her over a box, Darlene heard the contents tense. The snakes had felt the vibrations of their approach.

"You take the lid off the box," Glenn whispered. "Don't want my fingerprints all over it, do I." Even now, Glenn couldn't resist a brag about his fine plotting. Strangely enough, Darlene could see this was pretty smart for Glenn. It also struck her with a terrible, belated clarity that the only person he could confide in was the one who would not be able to incriminate him. Because he was truly going to kill her. Holding her head by the nape, he forced it towards the open box.

"You got a choice," he told her, ugly now. "Either you pick it up or you gonna get it in the face."

So Darlene had moved her left hand towards the rattlesnake.

~

In the afternoon, there came a radio announcement: the tornado warnings were being lifted, although a tornado *watch* remained in force for much of the Tennessee Valley.

The motel emptied as rapidly as it had filled.

"Looks like we escaped it this time," said the receptionist as the trucks lumbered on to the highway heading for Chattanooga and Huntsville, wet chrome gleaming. Her show of relief was unconvincing; that none of the trucks had ended up among the power lines, nor been put down among the trailer homes, scattering householders and household items across a fresh landscape of smithereened clapboard, was an evident disappointment to her.

The sun was shining by the time I checked out. The sodden earth was steaming, and crimson cardinal birds hopped about at the side of the road, beachcombing for dislodged or drowned insects. I crossed the river out of Scottsboro on the high suspension bridge whose great supports threw a semicircle of rusting iron tracery into the sky. The road climbed through beech woods towards Sand Mountain. Brown torrents, cataracts of mud and branches, poured down the road into the valley. My car laboured up the rapids and on to the plateau, where the road ran between chocolate-brown fields. At Rainsville, the biggest building in town belonged to the Church Pew Company. At Fort Payne, the sign outside Burger King commanded "Eat here—or we both starve." Preferring to chance it, I pushed on.

A thicket of signs selling fireworks, the result of some bizarre state ordinance, announced Georgia, afloat with pink cherry blossom. Prosperous Baptist churches had neoclassical porticos and a nice line in motoring metaphor. "Make the Lord your steering wheel, not your spare tire," exhorted one. "There are two finishes for cars; lacquer and liquor," warned another.

Then the road ducked deep into the woods, where Kingston lay in a somnolent hollow. The nearby highway had leached the life from the township, leaving empty rooms behind handsome red-brick facades, cardboard panes where the glass had gone and the passing of an occasional freight train that might once have halted here. The businesses had long since closed, or relocated to the strip at nearby Cartersville. Neglect had settled in, running licks of rust down the clapboard of Kingston's bungalows and gradually breaking up the picket fences that enclosed the gardens. Even the plastic flamingoes had fallen over, or been repeatedly hit and run over by kids' bikes.

Only the Church of the Lord Jesus Christ, perched on a green knoll just outside of town, was neat—strikingly so.

Church members were gathering for the evening service. In the basement of the church, the women had laid out a pig roast, dishes of beans and salad and bottles of cola. Most of the men had finished eating, and were gathered deep in talk around their parked pick-ups. They wore slacks and shirts but mostly no ties; the collar button was undone to reveal a white V of working man's vest. Cuffs were ironed. Their shoes were polished. Their hair was cut short and kept tidy. They dressed with a scrubbed respectability that at once suggested decency, informality and modesty, and so placed them in another time.

Mostly, the women did not join them. When they had finished clearing away, they emerged, blinking against the evening light and palming away the gnats, to mix with the other women and their children. They were dressed in long, plain skirts, wore their hair straight and uncut, and did not wear make-up. Like their menfolk, they were gaunt people. Sallow skin stretched over angular jawlines, suggesting adversity, poverty and lingering illness. In many cases, their teeth were poor. They looked like austere innocents, subjects in grainy photographs: they looked like history.

It was the men who carried the boxes. The boxes were shallow, the size of briefcases. But they had been designed to be carried flat: neat handles protruded from the lids in the manner of tool boxes. The cheap ply or pine cut-offs from which they were made did not conceal the quality of workmanship that was almost devotional. The boxes were varnished, and encrusted with decorative metal studs to form crosses. On some of them, the words "Lord Jesus" had been stencilled in black. Gauze-covered squares had been cut into the lids like windows. They had heavy hasps, from which padlocks hung. *Steady*, I said to myself, guessing what they must contain.

I was wondering where the late twentieth century had gone when a gangly man in glasses approached me. Brother Bryce was from North Carolina, and Brother Bryce had the story of his life to share with me. He had only recently found God. Before that, he conceded, he had committed a "whole lot of sin."

"I drank too much," he told me. "I took too many drugs, and sure sold a lot more than what I took. I was in hell. Hell, I even sent people to hell. Then I woke up one night and there was a voice calling me back. I think I'd gone as far as I could, and now was time to get right with God. But finding the right church took some time. Tried the Baptists, Pentecostalists, tried them all. What I needed was something deeper than what most churches could offer. Hallelujah, so help me I found it right here, among these serpent handlers."

If I understood him right, Bryce had just told me he'd taken drugs, sold drugs, killed people and then found God. And we had only just met. His confession suggested that religion did not so much define the prevailing moral landscape of the southern Bible belt as the reaction to its seedy and largely unholy contours. The South had the Lord because it needed Him bad. I had noticed just how many churches there were in the South: now I understood it was for the same reason that the South had so many prisons.

At its heart, it struck me, Bryce's story was that of the prodigal redeemed. Before things went bad for him, Glenn Summerford had come to God this way. So too had snake-handling's founder, George Hensley, who had run with Tennessee's moonshine gangs until the Lord came to him in a vision on a mountain. By their own admission, many of those who were drawn to Holiness were in serious need of redemption. Theirs was a comprehensively dangerous world. These boys had left the spiritual offenses far behind. Hell, they had broken most of the federal laws too. First-degree hell-raising and drug-running had been more their style. They had reached about as low as you could get.

Getting right with the Lord, after a life of crime, was a characteristic of the twentieth-century Holiness experience; indeed, it was almost the defining signature of the snake-handlers' journey to faith and redemption. It was a story, I realized, I had

heard before: the story of St. Paul. St. Paul had stooped low,
persecuting early Christians, but had been redeemed. For many
Holiness men, not short of criminal convictions garnered in a
less holy past, St. Paul's example was the biblical lifeline that
convinced them salvation might yet be attainable, even by
them. In the Holiness churches, there was a particular empathy
with St. Paul. He had been where they had been, they liked to
think. They knew his life as revealed in the Scriptures. Most
particularly, they knew Acts 28, verses 5–6, where the ship-
wrecked St. Paul is bitten by a viper concealed among firewood:

> And he shook off the beast into the fire, and felt no harm.
> Howbeit they looked when he should have swollen, or fallen
> down dead suddenly; but after they had looked a great while
> and saw no harm come to him, they changed their minds,
> and said that he was a god.

Snake-handling delivered an instant assessment on their
spiritual condition: handling without incident confirmed them
to be true believers, like St. Paul. It was the thrice-weekly mir-
acle that authenticated the fact of their redemption. The Lord,
then, was in the serpent that did not strike. But He was also in
the bite that did not kill, whether it was St. Paul or Junior
McCormick that survived it. Junior, a short, rotund man with a
final fuzz of soft white hair keeping him from baldness, had
handled thousands of snakes in his time and had taken a bite
from a copperhead at a service only last week.

In his *Letters from an American Farmer* (1782), settler and
traveller Hector St. John De Crevecoeur memorably described
the symptoms of a bite from a copperhead, also known as the
"pilot" for its tendency to emerge from hibernation a few days
before the rattlesnake. "The poor wretch instantly swelled in a
most dreadful manner," wrote De Crevecoeur.

> A multitude of spots of different hues alternately appeared
> and vanished, on different parts of his body; his eyes were
> filled with madness and rage, he cast them on all present

with the most vindictive looks; he thrust out his tongue as
the snakes do; he hissed through his teeth with inconceivable
strength, and became an object of terror to all by-standers...
when in the space of two hours, death relieved the poor
wretch of his struggles, and the spectators from their appre-
hensions.

Junior looked like he had got off lightly. A thick medallion
scab was embossed upon the base of his forefinger. As he
showed the bite to me outside the church, the other men gathered about him reverently as if he had been touched by God.

"The Lord commanded us to handle snakes," said Junior with
a smile. "Didn't say nothing about there not being a few bites
along the way." It sounded like a well-worn platitude, the sort
the special status of bite survivor allowed him to make. I asked
him what it felt like, handling snakes.

Junior was silent for a moment. "Joy unspeakable," he pronounced finally. "To be holding death in your hands..."

The band members arrived with their guitars, drums, amps
and keyboard, and were soon setting up inside. Billy Summerford, Glenn's boy from an earlier marriage, was there. So was
Gene Sherbert, and Punkin Brown, whose wife Melissa had
died in Kentucky the year before. Gradually, the crowd picked
up their snake-boxes and followed the preacher, whose name
was Brother Carl, inside the church. It was a square room with
painted breeze-block walls and ceiling strip lighting that suggested a strict, functional simplicity. Decorative touches were
few, just ornate fake-brass ceiling fans and a Byzantine-style
arched window; the sort of room in which you might expect
your car to be serviced.

The band had arranged itself on chairs along the rear wall
beyond a low, wooden rail, and were idly plucking off guitar
notes. Between them and the ranks of pews stood a simple,
blond wood altar where the men set down their snake-boxes,
arranging them in a neat pile on the floor. As the crowd filed in,

Brother Carl leaned upon the altar, mouthing words like an amenable goldfish. "Praise the Lord," he said in his introspective and reedy voice. No firebrand, it seemed, was the small and balding Brother Carl. He wore brown clothes, had deep-set, inoffensive eyes and a putty face. Brother Carl had come to the Lord after years as a long-distance truck driver. The original white-painted clapboard building had recently been rebuilt in brick. Brother Carl was mighty proud of the new church.

Most of the gathering formed a rough semicircle around the altar. They steered clear of the pews and their grid rigidity. Holiness, a cry from the heart, an emotional reaction against chillier Presbyterian rites of service, had never been about pews. Pews contained souls who wished only to lose themselves in the Lord. I was an outsider, and English, so I took my place in one and felt awkward. A casual drum-roll sounded out over the low burr of conversation.

"So," said Brother Carl finally, checking his cuff buttons. "Let's have some music."

Suddenly, there was cacophony, and the balmy evening beyond the windows fled at the frenzy unloosed in its midst. The building shook to the combined rhythms of gospel and bluegrass, boogie woogie and rockabilly. Piety and ritual were absent from the music; passion, love and unrestrained expression reigned. It had an instantaneous effect on the assembly. They were no natural extroverts but within seconds they were clapping to a rolling rhythm of drums and guitar that sounded out across the car park towards the woods and the highway. Women and children grabbed tambourines and cymbals. The men tapped their toes and began to sway. As they reached into the music, they drew confidence from it, and they shed all that protective shyness they wore outside the church, in the wicked world.

I watched from my pew as they sang of the House of the Lord ("you'd better get in, into the House of the Lord"), of seeing glory in the mulberry tree, and of Satan being defeated ("the Lord still answers prayers"). One of the men grabbed a microphone and screamed into it, the deltaic veins flood-watering up his neck. The mood rose unstoppably. Everybody was dancing now, but not with any rhythmic grace. The movements had become manic, string-tight. Eyes were screwed shut or staring into space. Hands, raised in surrender, shimmied towards heaven. Some, unsure of their feet, were steadied by friends. Within minutes, these taciturn types had been transformed into unrestrained celebrants. They were letting the spirit take them, and they seemed to be going unconditionally. I looked on, an awed witness, and held tight to the pew in front of me.

Suddenly, a certain, unstoppable momentum seemed to have been achieved, and one of the men moved towards the boxes.

"You've got to know it's what the Lord wants you to do," Carl Porter would tell me later that evening. "Before reaching for snakes, you gotta wait on instruction from God. Either, you can receive instruction direct, when God will tell you the time is right, or you get the anointing. That's a kind of holy trance in which you're ready for anything."

The man rose from behind the altar, a large rattlesnake clasped in his right hand and an expression of wide-eyed abandonment rupturing his features I gasped and my heart began to gallop: I was among snakes. The man held the rattlesnake around the middle and rested its head upon the outstretched palm of his free hand. The snake's tongue flickered, but it seemed quite untroubled. For a while, the man stood still. He observed the snake while other men looked on, singing it down and vanquishing the devil with their joy. Then the music swept them all away, and they joyfully began to plunder the boxes and pass snakes among themselves like looted gifts. They scarfed them around their shoulders, or held them in their hands, plaiting them. One man hopped around on one leg, jiggling handfuls of snakes by his sides. Another man stood apart. He held three rattlesnakes in his hand and ran his other hand down

them, smoothing them out as if they were the contents of his
tie-rack. A melée of men stood trellised in rattlesnakes and copperheads. A man threw a snake in the air and caught it upon its
descent. Another arranged a small rattlesnake upon his head.
He caught it before his face as it fell and passed it to waiting
hands. One man held a snake in each hand and bounced across
the room, shouting "Praise Him!" A pale young man with
cropped hair, a rare tie-wearer, stood motionless, his right hand
clenched high above his head like Liberty. But his right fist was
closed around a handful of snakes, his eyes screwed shut, a
copperhead was stuffed in his breast pocket and several snakes
spilled from his trouser pockets. He was Liberty gone seriously
awry, a Liberty that slipped off his shoes and walked upon a rattlesnake when another man placed it at his feet.

Now, the women were beginning to handle. One stood quite
still. She had framed her face in a rattlesnake, holding its head
and tail behind her ears so it resembled a beautiful hairband.
Another, who had worked herself into a trance, approached the
altar, her arms outstretched and speaking in tongues.

"Mashala mikoya toala mishala," she gibbered. "Shalama
shala!" She was short of breath and close to tears. A woman
took her by the hands and cradled her head gently as if to comfort her: *There there dear.* At the other end of my pew, a woman
studiously changed a baby's diaper. Thank God, I thought,
something normal, as if you saw soiled diapers changed every
day in the churches I knew.

The mood finally crescendoed, reaching a peak where it
could climb no further. A kind of rapturous exhaustion had
crept over the gathering. The music slowed, the frenzy ebbed
and people returned from the entranced places they had been.
The snakes were returned to their boxes. In the quiet, you
could once more hear the comforting sounds of evening: cars
passing on the road, a crow calling and wind in the sentinel
oaks.

Later, people stood up unannounced. They called it testifying, speaking of their belief and also of those who needed their prayers. Some spoke briefly. For one man, merely telling us that the Lord had shown him a safe road home in the mist just the other night seemed to suffice.

"God's with us whether we're driving home, doing dishes or handling serpents," he then concluded. "I want you to pray for me tonight." Around him, cymbals clashed and tambourines rattled in approval. Others rose to speak at greater length.

"There's people tell me I'm living it strange," the young man with the cropped hair and the tie began, "just 'cos I sometimes feel I ought to tread upon serpents, and maybe upon the scorpion, hallelujah, and upon the devil, the enemy of the people." And what, asked the affront in his voice, was strange in that?

"And when people ask," he continued, "how can I handle them serpents, I tell them of the young eagle in his nest, hallelujah, who don't yet know what he's gonna do. What he does know is that mama can fly. Sometimes, she leads him to the edge and then she takes him back to safety, and it might be the first time, it might be the second, it might be the third, but the time comes he finally thinks I've got one of these and one of these," he was flapping an arm, then the other, "and what do you know? He's flying."

A special welcome was reserved for Junior McCormick, the church's most recent bite survivor, who also had something to say about flying. Junior's family had been urging him not to handle snakes since he might get hurt again.

"Since I might get hurt," Junior echoed his family resignedly. "Well, you might stay at home and an airplane sail down through your house." An F-14 had done just that to a Nashville suburb only a few days earlier, killing five people.

"But you know," Junior continued, "it don't matter whether mama likes it, daddy likes it, your children like it, but you just gotta do the word of God. Ay-men. Praise God. Because you know how the Bible says heaven and earth shall pass away, and one day he's going to call us from this walk of life, and it don't matter about a few lil' ol' bites along the way. Ay-men." Just as he was about to sit down, Junior remembered something. "These folks dancing round here," he remarked, jabbing a finger at nobody in particular, "they got something special and it didn't come out of no whiskey bottle."

Brother Carl suggested prayers. In a final collective apoplexy, the members threw themselves to the floor, banging their fists and uttering high-pitched whines that filled the church with the sound of a motorcross rally. And then the service was at an end, and they emerged from the church in an orderly shuffle, as if nothing had happened to them in the few hours past.

"People call us all kind of things," said Brother Carl, reflective now, as we stood outside under a newly risen moon, watching the gathering disperse. The service had been something, even by Holiness standards; Brother Carl could see how an outsider might wonder. "But we're just obeying the word of the Lord," he said. "I don't rightly see but we ain't got us no choice but to do what it says in the Bible. And if that means picking up serpents, well, I ain't the one to deny Him."

"But what if people are bitten and get sick? What does that mean?"

"We call that getting ahead of the Lord. You've got to listen to the Lord. Perhaps you've tried to handle before you're ready. Perhaps your faith's not perfect. Now, that don't make you a bad man or nothing."

It troubled me, I said, that handlers sometimes died from snakebite. That family and friends let handlers die.

"Well, sometimes there's things we can't understand," Carl reasoned. "It may be the sufferer's time has come, may be that the Lord has called him home. I'm for believing in personal preference. If a person who's bit wants to go to the hospital, I'll take him right there. But if a person wants to trust in the Lord, I'm not the one to put a doctor between them."

I asked Brother Carl what he thought of snakes.

"There's a mean spirit in them," he replied, "a spirit inspired by the devil. When the anointing's on me, the snakes can't touch me. I'm safe then. But when I'm alone, I'm real scared of the things." He was using the word "alone," I guessed, to mean outside a state of holiness; alone without the Lord. Which was when snakes stalked the earth, sin incarnate.

Brother Carl's words reminded me how Christianity had always had a particular hostility towards snakes. Apart from the odd significant dove, animals are generally restricted to minor appearances in the Bible. The glaring exception is the snake, which is allotted the most unenviable role going. That it should be the serpent that by its subtlety robs man of immortality implies an extraordinary aversion towards this creature, and one that self-evidently seems to have flourished in modern times. Sin and the Serpent, with all its endorsing, alliterative sibilance, is perhaps irresistible anthropomorphism, animal and attitude making a perfect match.

While the Christian mainstream tended to view the serpent's subtlety rather more figuratively, Brother Carl's snake-handlers were unique in that they demonized the real thing. They saw the devil embodied in living snakes. They reminded me of a missionary called Elijah Bingham. Bingham, who had worked in Nigeria in the 1960s, published a pamphlet called "Snake Stories from Africa," in which he drew parallels, often tortuous ones, between the nature of snakes and sin. From every snake he encountered, the obsessive Bingham drew platitudinous Christian lessons. "Often at night, especially in the early rainy season," he wrote, "we have almost trodden on little snakes in our path...there are little snakes in your path and mine daily, ready to bite, and some bites may be fatal (at least to our Christian walk)."

Bingham saw snakes as the messengers of "Satan, that big serpent," sent "to bite us and inject the venom of sin." He drew Christian parallels when observing a charmer remove a set of snake fangs: "How like the Lord Jesus, I thought. He has taken away the sting of death for all believers." And he even saw the

devil's hand in the way snakes soon grew new fangs: "If you
or I," he warned, "neglect to watch and pray we may find our
sins taking root again." The answer, of course, was light, or, as
Bingham intended to imply, The Light. "We have carried a
torch or a lantern," he concluded triumphantly, "and dazzled
the snake with the light till someone killed it."

The car park was empty now. The last of the departing headlights had corkscrewed into the dark.

"It's turning warm tonight," said Brother Carl. "You watch
out for them snakes, mind; they like the warmth lingering on
the asphalt." He shook my hand and returned to the church to
lock up. I could have done without the mention of snakes. I
didn't believe snakes to be agents of Satan but I feared them
nevertheless, so much so that I was convinced I would tread
upon any which happened to be about. I fished from my pocket
the torch I always carried in snakeland, snapped it on and
shuffled forward into the sanctuary of the beam. Every few
steps, I stopped and turned, shining the torch all around me to
check my surroundings, and in so doing created a momentary
pattern that was almost emblematic; a long line of light on
which fast-fading searchlight circles were continually being
threaded. There was comfort in these circles. To me, they felt
like protective charms that outlasted the momentary lightfalls
that shaped them. I stepping-stoned my way across Brother
Carl's car park without incident, and drove into the night, wondering at what I had witnessed that evening.

Perfect victory, the handlers called it. But not always. There
were pretty big defeats out there too. Punkin Brown knew that;
he'd put his wife, Melissa, in the ground only last fall. What
could they do in the face of such deaths but trust the Lord's
purpose? It must, it struck me, take an awesome amount of
faith. As if to justify snake-handling, Junior McCormick had
pointed out that an airplane of the kind that fell out of the sky
on five Nashville folks could get into trouble any time, in a
patch of sky near you. You could die anyways was his drift. But

being buried by an F-14, that was bad luck. That wasn't court-
ing trouble. Not like handling snakes was. Snakes bit people.
Not always, but biting was what snakes did.

On the radio, a report told how a boy had died that day in
Mobile, Alabama. Playing Russian Roulette.

"I have heard of their having been seen, formerly, at the first
settling of Georgia, seven, eight and even ten feet in length." So
wrote the anonymous author of the pamphlet *Wonderful
Account of the Rattlesnake and Other Serpents of America*, in
1805. He was stopped on a Georgia footpath, he wrote, by "the
sight of a hedious serpent, the formidable Rattle Snake, in a
high spiral coil, forming a circular mound half the height of my
knees, within six inches of the narrow path." He described the
typical rattlesnake's "constricted lips," his "mortal fangs," "his
eyes red as burning coals," and how "his brandishing forked
tongue of the colour of the hottest flame, continually menaces
death and destruction, yet never strikes unless sure of his mark."

Early settlers from snake-paltry Britain were impressed by
the rattlesnake. Some even claimed it could fly; others, that it
could poison merely with its breath. During his American travels in the late eighteenth century, the French poet and mem-
oirist Chateaubriand heard of a snake of the South that
poisoned "the atmosphere which surrounds it. It decomposes
the air, which, imprudently inhaled, induces languour. The per
son wastes away, the lungs are affected, and in the course of
four months he dies of consumption." Sir Thomas Walduck, addressing the Royal Society in 1714, reported a similar effect. In
the proximity of rattlesnakes, said Walduck, "woodsmen...are
in a fright as tho' a Spectre was near them, and that the snake's
breath inflames ye Air and before they either hear or see them
they are seized with sorrow."

Corrupted air supply worried some settlers, but most settlers
mainly feared the bite of snakes. In a 1634 pamphlet devoted to
describing the "Commodities and Discommodities" of New

England, a Dr. Higgeson wrote, very much in the Discommodities section: "Yea there are some Serpents called Rattle Snakes
that have Rattles in their Tayles, that will...sting him so mor-
tally that he will die within a quarter of an hour after..." "It is
certain," wrote the anonymous pamphleteer of 1805, "that he is
capable by a puncture or scratch of one of his fangs not only to
kill the largest animal in America, and that in a few minutes
time but to turn the whole body into corruption."

De Crevecoeur, who devoted an entire chapter of his highly
successful *Letters* to snakes (and also, without apparent logic,
to the humming-bird), wrote of the Dutch farmer who was
struck at by a rattlesnake when he "went to mowing." That
night, "the farmer pulled off his boots and went to bed; and was
soon after attacked with a strange sickness at his stomach; he
swelled, and before a physician could be sent for, died." A few
days later, the son pulled on his father's boots, similarly sickened and died. In time, "one of the neighbours, who bought the
boots, presently put them on, and was attacked in the same
manner as the other two had been." It was only then that an
alert doctor found the fangs from the original strike embedded
in one of the boots, and so put a timely end to a series of deaths
that was otherwise destined to last as long as the boots. It's a
story, as commentators have noted, with a popular modern version in which Texas garage mechanics attempting to change a
punctured tire keep dying, for no apparent reason...

The symptoms of rattlesnake envenomization were apparently no less colourful. Thomas Lechford in 1642 described a
bite victim turning "blew, white and greene spotted." From
these accounts it is not hard to see why the timber rattler—as
common in the Appalachian woods as the diamondback is on
the lowlands to the south—was named *Crotalus horridus,* or the
horrid rattlesnake; on a par, in terms of taxonomist hysteria,
with *Ursus horribilis* (horrible bear), that other scourge of the
American wilderness, the grizzly.

Some, of course, rejected such accounts of snakes as arrant

nonsense. Even in 1654, when the existence of dragons, griffons and seahorses was accepted elsewhere, the level-headed Thomas Morton wrote in a treatise about New England that, "It is simplicity [simple-mindedness] in any one that shall tell a bugbeare tale of horrible, or terrible serpents, that are in that land." Even De Crevecoeur conceded rattlesnakes could be "extremely inactive, and if not touched, are perfectly inoffensive."

But it was the pointed implication behind the words "if not touched"—that the rattlesnake was impressively equipped to defend itself if so required—that seemed to appeal to the symbolically minded among the American liberationists, thoroughly disenchanted by excessive English taxation demands. For the rattlesnake and the Americans of the 1770s seemed to have much in common.

The rattlesnake, the *Pennsylvania Journal* wrote in December 1775,

> may be esteemed an emblem of vigilance. She never begins an attack, or, when once engaged, never surrenders. She is therefore an emblem of magnanimity and true courage...She never wounds until she has generously given notice even to her enemy, and cautioned...against the danger of treading upon her.

Few Pennsylvanians were fooled into reading the *Journal* piece as a natural history article. The newspaper was actually offering a thinly veiled statement of American resolve in the face of war with England. For the rattlesnake, which perfectly symbolized the prevailing mood among Americans—a stout determination to defend themselves, and to defend themselves ferociously—had emerged as the unlikely mascot of the fledgling nation.

"The colours of the American fleet," the alarmed *London Chronicle* reported on July 27, 1776, "have a snake with thirteen rattles...described in the attitude of going to strike, with this motto, 'Don't tread on me.' Such a flag was presented to Congress in February 1776. It was placed to the left of the

President's chair, and flew at Trenton, New Jersey, in the same year, where George Washington defeated the British.

Long before the adoption of the Stars and Stripes, the rattlesnake had stood for the spirit of liberationist America, defiant, fearless and quite prepared to defend itself. By 1996, however, America had moved on and its opinion of itself had changed. America had disclaimed the rattlesnake. The snake had suffered a fall from grace. It had become mean, malevolent and it now belonged to Satan.

6. AUSTRALIA II

All of which reminded me, as my connecting plane headed north from Sydney in the musing hours of the morning, of a shred of history that had long intrigued me: to the city disappearing beneath me among spring clouds, the British had dispatched a ship in December 1846 called HMS *Rattlesnake*.

HMS *how so?* Most ships, particularly British ships, got names like *Victory, Resolution* or *Endeavour,* names that redounded to the honour and glory of the Empire. In the circumstances, it seemed to me that this one might as well have been called *Bannockburn* or *Battle of Hastings* or, in another age entirely, *Mad Cow* or *Warm Beer* or *Carlton Palmer* or some other name pertaining to British ignominy. Was this the work, I wondered, of some fifth-column humorist in a position of some influence on the Royal Navy's vessel-naming committee? Who soberly proposed as a name the *Rattlesnake* while privately exulting at the notion of a maritime billboard to commemorate Britain's American humiliation of seventy years before? As she barrelled down the Solent on a squally winter day, the 28-gun *Rattlesnake* was a reminder to the world how the self-reliant Americans had rallied behind the rattlesnake motif to gain their freedom and rob Britain of her prime overseas possession. Was sending the *Rattlesnake* to Australia—to evoke pompous British generals in powdered periwigs being put to flight by brave and honest hayseeds inspired by the righteousness of their cause—really the best way of flying the flag?

Australia was perhaps unready for independence in 1846, but the colony was certainly growing up fast. There was, moreover,

no shortage of nationalist sentiment abroad, and none would
have better appreciated the rich irony in the *Rattlesnake*'s name
than William Charles Wentworth. Wentworth, whose mother
had been transported from England for theft and whose father
had been charged with highway robbery, had led the recent
campaign against the hated Ralph Darling, the English Governor of New South Wales. Wentworth was also responsible for
New South Wales's constitution, coined the slogan "Australia
First," and founded the newspaper, the *Australian*. He champi-
oned the ideal of the self-reliant Australian, so challenging the
stereotype of the inveterate skulker and mean petty thief,
shackled forever to his criminal origins.

Wentworth had succeeded Sir Henry Browne Hayes, long
since pardoned and permitted to return to his beloved, snake-
free Ireland, at Vaucluse House and turned the errant knight's
"snug cottages," as a local newspaper described it, into a handsome plantation house. After six months at sea, the land-
starved *Rattlesnake*'s 190 officers and men looked out on "a
splendid and unequalled Marine Estate, commanding a perfect
view of the head lands and sinuosities of the various bays and
inlets of the truly romantic Harbour in which it is situated,"
as a contemporary newspaper advertisement had described
Vaucluse when it was put up for sale. It was in these same gardens that Wentworth had thrown a lavish party in 1831 to celebrate the hated Darling's departure for England. That night,
4,000 guests had watched an impressive firework display which
emblazoned the message DOWN WITH THE TYRANT across
the sky as Darling's ship disappeared towards the horizon.
Doubtless, when the *Rattlesnake* appeared off Sydney on July
16, 1847, Australia's first nationalist would have permitted
himself an amused chuckle and perhaps imagined the day that
Australia would unshackle herself from Britain, and finally fly
her own flag. Not that any snake would feature on it.

Early Australians may have seen the American independence
experience as an example to all young nations, but few of them

could understand the reptilian empathy that had put a rattlesnake on the American flag. For while American snakes were largely considered a hazard only in the field—"the most danger of being bit by these snakes is for those that survey Land in Carolina," as one eighteenth-century pamphleteer put it—Australian snakes were numerous enough to be regularly described as "verminous" or "infestations" and, as Browne Hayes had discovered, they regularly appeared indoors. Moreover, they were widely acknowledged as being considerably more deadly than anything encountered in the Americas.

Early Australian journals and diaries crawl with snake sightings. On August 9, 1849, Sydney resident Henrietta Heathorn wrote of a typical encounter when walking in the bush with her friend, Alice Radford and her half-sister, Oriana.

> We had not gone far in the bush when Alice screamed "a snake," seizing hold of Ory who immediately sprang to the other side. I looked around for a stick and killed the creature, breaking two or three brittle ones over him. It was of a lead colour and about three feet long, a deadly one...I was very brave and I looked upon myself as a miniature heroine.

By this time, Henrietta Heathorn had already met the man she would marry, a young officer from the visiting British survey ship, the *Rattlesnake*. Thomas Huxley was the ship's assistant surgeon, but in time he would achieve international renown for his work on evolutionary theory. On board, he took a pronounced interest in the investigations of the ship's official naturalist, John MacGillivray. To these two men at least, the *Rattlesnake*'s name made excellent sense, harmonizing as it did with their scientific and zoological interests.

MacGillivray and Huxley's enthusiasm for the new continent's unique fauna and flora was consistent with a generally more enlightened attitude towards Australia then emerging. Although the *Rattlesnake*'s southerly course to the Antipodes, via South Africa and the Indian Ocean, echoed that of the transports, she travelled in quite a different spirit. Despite the fact that convicts would continue to be off-loaded in Australia for another twenty years, transport volumes had long since peaked and by the 1840s the colony was fast outgrowing its original, unenviable function as dump chute for Britain's undesirables. "Free emigrants" lured by cheap land and the promise of opportunity were heading for Sydney and the fledgling settlements at Melbourne and Adelaide in ever-increasing numbers. It is to be remembered that John Girling, the unfortunate reptile attendant at London Zoo, had been seeing off a friend bound for Australia before his fatal encounter with a cobra one morning in 1852.

The Admiralty had instructed Owen Stanley, captain of the *Rattlesnake,* to land at the Cape of Good Hope en route to Australia so as to take on fresh water and deliver £50,000 to the colony, and also at Mauritius to "land the treasure there" (£1,500). But the ship was more than imperial delivery girl. The crew had also been entrusted with a number of chronometers which they were to verify for accuracy on the initial leg to Madeira. Upon arrival in Australia, they were to chart the continent's little-known eastern coast—fast emerging as an alternative, albeit reef-strewn route to and from the Antipodes, via the Torres Strait and India. Captain Stanley was to explore the inner passage that separated the Queensland coast from the Great Barrier Reef, dispatch landing parties to assess the potential of future harbour sites, and to "determine which was the best opening that those reefs would afford, and to make such a survey thereof as would ensure the safety of all vessels."

By April 1848, when the *Rattlesnake* left Sydney for a ten-month absence surveying the Queensland coast, Thomas Huxley was hopelessly love-struck, and spent much of his time penning letters to Henrietta Heathorn. But MacGillivray suffered no such distractions. The *Rattlesnake* was sailing north into a world as unexplored as Botany Bay had been sixty years before, when the First Fleet had poked its provisional bows into those dark, unwelcoming shallows. North of Moreton Bay, latterly Brisbane, where a particularly inhumane penal settlement had existed for twenty years, the map was all but uncharted. MacGillivray saw "butterflies of great size and splendour, with dark purple wings, broadly margined with ultramarine." He collected extensive specimens such as pheasant-tailed pigeons, humming-birds and bustards, stingrays and crabs. The crew piled up cowries, cones and spider shells along the beaches, like children in a prelapsarian paradise, and MacGillivray's entrancement was complete.

MacGillivray also noticed the snakes. "Snakes," he wrote in his journal, "require to be carefully avoided. One day, I killed single individuals of two kinds—one a slender, very active green whip snake, four feet in length—the other, the brown snake of New South Wales, where its bite is considered fatal."

On July 26, 1848, the *Rattlesnake* anchored off Trinity Bay, some 1,500 miles north of Sydney. The landing party that left before daylight found a "wide creek running through low mangrove swamps and with the eye could trace its windings for two or three miles."

So could I. My view was from the air as the plane banked over the Coral Sea, grinding and bumping between thick, tropical clouds like a jostled dancer on a crowded disco floor. The flight from Sydney, which had traced the voyage of the *Rattlesnake* northwards, was almost at its end above that same creek that Owen Stanley had adjudged "useless" one hundred and fifty years before. The creek shuffled like a fat, grey snake into Trinity Bay south of a town whose apparent size belied Captain Stanley's pessimism. As the plane approached Cairns airport, we were low enough to see streams of cars flashing in the sun as they beetled along the promenade, the thicket of masts in the marina, the hotels, shopping malls and the sprawl of suburbs nibbling at the bush which hemmed in the town on the landside.

From the air, the Cairns landscape had seemed vigorously coloured. The blue sea relieved the sombre greens of the bush and the murky brown shallows of the mangrove swamps. At ground level, however, the sunlight was a harsh wash that stripped the landscape of its varied tones, and the heat came in damp, tropic gusts that caused clothes to wilt and raised an immediate rim of sweat beads along my hairline.

I drove into town, a grid of malls and arcades, car rental firms and tour operators, motels and bars. In the window of a souvenir shop I glimpsed a yellow road sign featuring a wide-jawed crocodile in a triangle above the words "I love tourists." I wondered whether the sign was all beery good humour or whether it reflected a genuine local indifference, perhaps even a disdain, towards the tourists who flocked here in ever-increasing numbers. In FNQ, popular acronym for Far North Queensland, you could snorkel and dive the Barrier Reef or overland the 600 miles of dirt road to Cape York at the northernmost tip of Australia. You could take day tours into the Daintree, one of the oldest forests on earth, ride the train up to Kuranda in the hills above the town, raft white water on the Tully River, go hot-air ballooning, bush-walking or even bungee jump. There was no doubting that tourists loved FNQ but it seemed the locals were less ready to reciprocate, preferring to let their crocs do the loving in a carnivorous, if figurative, sort of way.

Cairns was good at tourism; it just wasn't sure that it wanted to be. Hell, it had always been an honest, hard-working frontier kind of place. Had they really hacked out the mangroves to settle this place so the townsfolk could serve pizza to moneyed-up strangers? From what I was hearing on the radio, however, one thing was clear; tourism had not yet had the least sanitizing effect upon the town's frontier manners. On the afternoon show, people who had been sick over strangers at concerts were being invited to phone in with their technicolour reminiscences. The sheer number of enthusiastic callers soon persuaded me to steer clear of Cairns's live-music scene.

In a reading of the town's history, you cannot help but sense the people of Cairns being cheerfully sick over each other, just off-page and quite regularly, ever since the first settlement of the town in the 1870s. A whiff of old vomit, as well as alcohol, gunpowder and cheap rouge, pervades the annals of a town that grew up to service the gold strikes inland on the Hodgkinson River. Gaming saloons, grog houses and brothels proliferated. The newspaper at comparatively genteel Cooktown a few hundred miles to the north scoffed that Cairns could never be anything more "but a pack road to the Hodgkinson from Trinity Bay and therefore... worshippers of Bacchus will be the sole inhabitants of death's stronghold, the new port, for years."

Early Cairns was certainly no place to leave your boat unattended on the beach. Owners often returned to find them blown to pieces by the uncertain but persistent musketry of drunks who had mistaken the crafts for crocodiles. Citizens who took a stroll under the gaslights along the promenade were in danger of being run down by ships; drunken mariners tended to mistake the newfangled illuminations for port entrance lights, and regularly ran their vessels aground with a rending of wood and iron.

Predictably enough, broken boats and damaged nerves soon made Cairns something of a magnet for solicitors. The first to follow his intuitive nose north to new business was C. Edwin Smith in 1876, securing the best pitch in town long before services turned up that elsewhere might have been considered rather more essential, like hairdressers and chemists, for example. Smith and his kind were soon knee-deep in clients, and never more so than in April 1886, when the celebrations to mark the commencement of construction on the railway inland to Herberton went with a characteristic Cairns swing. The day began inauspiciously when the contractor failed to turn up with

the presentation wheelbarrow and spade, so preventing the
waiting premier of Queensland from turning the first, ceremonial sod of earth. From there, things went rapidly downhill. The
official banquet for 250 invited guests passed without mishap
but the barrels of beer and the roasted bullock that were set out
for public consumption soon led to trouble. When the town
lockup reached capacity, the authorities were compelled to put
other inebriates in an old punt on the creek, which soon threatened to capsize from sheer weight of numbers. "The scenes that
occurred in Abbott Street yesterday," said the town's enraged
police magistrate in the aftermath of the celebrations, "would
have disgraced a village of New Guinea cannibals or a black fel-
lows' corroboree."

I parked just off Abbott Street, near a shaded precinct of
shopping malls and cafés that leaked Muzak and sullen food
smells into the leaden, afternoon air. The face of a falling
drunk, who might have been here since April 1886, had just
embedded itself in a flower bed outside a café called The
Swagman. The drunk was being gently levered into an ambulance whose blue light revolved resignedly. The earth-moulded
impression of the drunk's booze-blind face stared up at me from
among vomit-splattered gladioli.

I sat under a parasol, reading the *Cairns Post*. "Barrier Reef
penile shock," read the unlikely headline. "Internet users seeking information on the Great Barrier Reef," it explained, "are
being confronted with lists of unrelated Gold Coast businesses,
including a clinic offering penis enlargements." Under the system, it continued, "people who type in the key words 'Great
Barrier Reef' don't quite get what they bargained for."

But then, this was Cairns. With a frontier history like that of
Cairns, what did net surfers expect? A sign in the window of a
pharmacy, where I went to buy some film, demonstrated the
point. "Buy a twin pack of Kodak film," it said. "Get a roll free
(Choose from our counter staff)." I glanced at the counter staff
and chose not to. On the pharmacy counter, I noticed a pack-

age addressed to "Catherine, Catering, Kidston Gold Mine," I did not envy Catherine her job, I decided.

The Cooktown newspaper may have been right to accuse Cairns of licentiousness, but its claims that the new settlement would never get off its knees were soon looking complacent. By the late 1880s, land values in Cairns were booming. In one typical sale, a five-acre site realized a record price before the buyer discovered that his land lay under the Coral Sea at high tide. (Cue C. Edwin Smith.) Goods were being landed and disembarked at a ferocious rate: so much so that the town pier collapsed under the weight of a coal delivery. Perhaps this was not good news for the town pier, but it was a sure sign of a thriving city.

Tourism soon began its inexorable rise. Nearby Green Island, whose restaurants and aquaria are now serviced by regular, high-speed launches, was established as a favoured shooting and fishing resort as early as the 1880s. The usual luggage for a weekend on the island was said to include Martini-Henry rifles, twenty charges of dynamite for fishing, sixteen jars of whisky and a bottle of brandy in case of snakebite.

Snakes were a constant danger. A medical man called Johnstone, who had lived for over twenty years in FNQ, was quoted in the *Australasian Medical Gazette* of 1890 as having seen "twelve people who had been bitten by brown snakes," and having "attended the funerals of the lot," as well as four deaths out of seven bites from death adders. Cases of snakebite received cursory coverage in the local paper; evidently, they were less hard news than an all-too-frequent fact of FNQ life. A report on the death from snakebite of Robert Crabbe, aged eleven, on October 13, 1920 was included in the small notices section of the *Cairns Post* alongside "Price of Groceries," "Gordonvale Caledonians" and "*Geranium* Crew to be Entertained." (Evidently, the *Rattlesnake* was not the only ship out there with a silly name.) "The lad," the report read, "accompanied by a brother, was crossing a piece of bush land on his way home about seven o'clock on Tuesday night when he trod on some-thing which bit him and, it is stated, the boys saw something crawling away…" Exactly a week later, Donald Means, aged eight, also died from snakebite in the Cairns area. "Dr. Clarke did everything possible to fight the venom, but the deadly poison had gone too far when the lad was brought to him," the *Post* reported. "A second case of snakebite occurred in the country on Wednesday," the report continued, "a young girl named Doris Aldridge, aged eleven years, residing at Kamma, being attacked by a reptile." "Yet another case of snakebite," the *Post* reported on November 15 that year: "A lad named Jackson, aged nine years, residing with his people at Aloomba, was bitten by a green snake on the left heel."

These early Australian accounts of snakebite are numerous, but they are also characterized by a persistent vagueness. Despite the regularity of snakebite, doctors, victims and journalists alike tended to refer to "reptiles," "green snakes," "black snakes," "lead-coloured snakes" and "brown snakes" with what sounds like a marked lack of species certainty. This was unsurprising. The pronounced physical signatures that tend to make snakes elsewhere easily distinguishable—most famously, the rattle of the eponymous American snake; the hood with the mark, like an old-fashioned pair of spectacles, of the Indian cobra; or the zig-zag markings of most vipers—are quite absent in Australian snakes. No Australian snakes rear up like the black mamba. No Australian snake spits like the African spitting cobra. And no Australian snake has anything like the patch-patterned, trash-couturier magnificence of the African gaboon viper's markings.

All, then, that is left to the average Australian to identify snakes is colour, a reliance that is strikingly reflected in the names of Australian snakes. Most commonly, there is the brown snake, which is brown. The red-bellied black snake has a red belly and is black. Gerard Krefft's *The Snakes of Australia* was published in Sydney in 1869. In the names of the snakes

Krefft details—orange-bellied brown snake, grey snake, scarlet-spotted snake, brown-banded snake, pale-headed snake, black-naped snake and black and white ringed snake—can be heard the echoes of the original settlers' untutored utterances as they breathlessly describe to friends and physicians the snakes they have just encountered—or received a bite from.

On occasion, however, Australian snakes do attract more exotic names. The name of the king brown seems inspired by, not to say blatantly lifted from, Southeast Asia's king cobra. Regal "king," however, changes in conjunction with mundane "brown," and this highly venomous snake ends up suggesting a forgettable species of butterfly. Or a soul artiste. The tiger snake sounds a little more adventurous but was actually named not for its ferocity but for the yellow stripes that can mark this species. And even the dramatic-sounding death adder was originally known as the deaf adder. An innocuous sounding name and, since all snakes are deaf, about as informative as calling a species of kangaroo the "bouncer." The corrective name change certainly put some belated pep into Australian snake names, but it still left this snake mistakenly labelled: the death adder is not actually a member of the adder, or viper, family.

So it is that you have to look elsewhere for all the best snake names; the imperious bushmaster or the wickedly invasive fer-de-lance of Latin America, the quasi-mystical black mamba or the perfectly named, malevolently tumescent puff adder of Africa, the evil-sounding krait of India, the menacing sidewinder or the darkly poetic water moccasin of North America. The Australian exception, widely regarded as the world's most dangerous snake, is the taipan.

In the case of this Queensland snake, larger, faster and generally more impressive than its Antipodean counterparts, a typically blunt Australian offering like "rusty brown snake" would simply not wash. The name "taipan" has been variously and uncertainly ascribed to the Wikmunkum aboriginals of remote Cape York and to Queensland canefield labourers from Malaya, where the word denotes "master" or "boss." Either way, its name has a dark, exotic resonance that leaves other Australian snakes for dead. Which, over the years, is what it has done to a number of Australians.

It was not until the years following the second World War, when the population along the Queensland littoral began to rise, that the taipan first invaded Australia's collective consciousness with a vengeance. The late 1940s and early 1950s saw a series of snakebite deaths in the region. In each case, it was noted that the victim had failed signally to respond to the available antivenoms, which were regarded as effective against the bites of all the commonly encountered Australian snakes.

The fatal bites had occurred far south of what was then accepted to be the taipan's highly restricted range: the largely uninhabited Cape York in what might be termed EFNQ or Extremely Far North Queensland. It was subsequently established however that taipans had indeed been responsible for the snakebite deaths. Which meant that taipans were actually resident in Cairns and even as far south as Brisbane. Suddenly, an unfamiliar snake with an exotic name was the talk of the Queensland suburbs. Australia was awakening to the sensational horror of this alien snake in her midst. Without antivenom, you didn't survive a taipan bite. And no antivenom existed. You could joke about browns and make up songs about tigers, but taipans were something else.

"Weather change brings snake warning," read the next day's newspaper. "A warning has been issued to all residents to stay away from snakes following the recent increase in sightings due to the weather becoming hotter and drier," explained the paper. Don't try to catch them, it continued. Keep an eye on the children. Keep the grass short around the house. Remove piles of rubbish from the yard. If a snake is found in the home, the paper advised, isolate it in a room and then call for a trained person to remove it.

Such a person was David Williams. If a snake turned up in Cairns, David was the man to call. He had been interested in snakes since his childhood. From the first opportunity, this self-confessed snake nut or "herp" had worked in snake parks down south, was currently involved in a snake venom-extraction project in Papua New Guinea, and had once spent twelve hours a day for two weeks penned in a glass cage with two brown snakes and three taipans at the top of the Sydney Tower to raise money for the Royal Flying Doctor Service.

"There was a telly outside the pen," said David. "So I caught up with the soaps."

Truly, David looked more like the sort of person you'd find slouched in front of a soap opera, a beer and peanuts in hand, than squatting in a snake-pit. He had joined me in a chic coffee bar above the marina, where we watched a trail of yachts and hydrofoils decorated with honey-coloured girls in bikinis heading for the reef. You could feel the contents of the coffee bar riling at our approach. David looked about as suitably fashionable as me, which was not very. He was overweight, and dressed with all the style of an actuary on a very rare holiday break. In the modish surroundings, he almost seemed anarchic. He wore glasses and spoke through his nose. His skin had the soft, sun-free pallor of a man who lived under artificial light.

Until recently, he told me, he had kept a large collection of snakes and other reptiles in a Cairns warehouse. Then the parks department had raided him, and charged him with failing to look after the animals properly. So Williams had put the collection on a truck and driven it upcountry to Ravenshoe, about an hour inland, where Brian Starkey agreed to look after them in the empty premises of Ravenshoe's former pet store. Brian, a friend and colleague from way back, also lived for snakes.

"At this moment," reflected David, grinning gleefully at the notion, "there's probably more taipans in the Ravenshoe pet store than in any one place anywhere else in the world. If only that town's good people knew."

I was interested, I told David, in the way the taipan had incited the Australian imagination to something close to panic.

"I guess the whole taipan business really took off with Kevin Budden back in 1950," he explained. "Here, finish your cappuccino, and I'll give you the Kev tour."

He led me to his car and drove out along the Esplanade. The tide was low. Green mangrove shoots were poking out of the mud. Further out, the harbour was thicketed by masts.

"Cairns Base Hospital to the left," said David. "Probably seen more taipan bites there than anywhere else in the world." He turned left towards Sheridan Street, the main road out of town.

"Kevin was a young herp up from Sydney," explained David. "He had come to Cairns in July 1950 looking for taipans. At that time, all the notable herps were after taipans, and Cairns was about the best place for finding them. Catching taipans meant a first supply of venom on which to start producing antivenom serums. Until then, nobody had bagged a live one. The prize was a place in herp history."

On the morning of July 27, 1950, Kevin walked out of town in search of taipans. "Somewhere, along here, Kevin turned left and crossed the railway line," said David, turning left and crossing the railway line. "At that time, much of this was still cane field or uncleared bush, so there were plenty of taipans about. He must have walked down here, passed the Botanical Gardens on his left and the old oil tanks on his right, and turned right into Friend Street," he explained, passing the old oil tanks on his right and turning right into Friend Street. Small suburban houses flanked the street on either side.

"All along the way," David continued, "Kevin was kicking at piles of rubbish, hoping for some evidence of snakes, and overturning old pieces of scrap iron, logs and branches. It was only when he got to Friend Street, to the house you see on the left, that he got lucky. He found a six-foot snake among a pile of rubble at the back, grabbed it around the neck and pulled it free."

Kevin was pretty sure he had just caught a taipan; the bad news was his grip was awkward. Reckoning he could not safely bag the snake on his own, he walked down to the main road. A passing truck driver suddenly found himself giving a lift to a man holding what he claimed to be Australia's most dangerous snake. "Take me to Stephens's place," the man with the snake said. "He'll be able to identify it." Dr. Stephens, a local snake expert, had seen enough dead snakes to confirm that Kevin's snake was indeed a taipan. It was as they were trying to bag the snake, however, that Kevin's precarious grip failed and the taipan bit him twice on the left hand. They got Kevin to the Cairns Base Hospital by eleven that morning where a tourniquet was applied. An hour later, it was dispensed with. They administered tiger-snake antivenom. The initial opinion was that Kevin had taken a mild bite; there were no convulsions or other apparent symptoms.

By three o'clock, however, Kevin was complaining of blurred vision. He was soon vomiting, began developing a headache and his pulse was up to 120 per minute. By seven o'clock, he could hardly move his tongue. Swallowing was very difficult. He could not speak, and communicated by pencil and paper. They gave him more antivenom. By eight o'clock, Kevin's face was paralyzed and he was struggling to breathe. They gave him a third dose of antivenom and put him on artificial respiration. He slept through the night but by nine o'clock in the morning his pulse was alarmingly weak. By midday, he was hardly breathing. He died at one-thirty in the afternoon.

Budden's death was reported nationally; certainly, such a story no longer had to compete for *Post* space with notices about entertaining the crews of latter-day *Geraniums*. By 1950, snakebite deaths had become a relative rarity and acquired a sensational aspect in the process. What made this story particularly compelling, however, was that the victim had successfully bagged the world's first live taipan before admitting himself to hospital. His death, in the heroic phrase, may not have been in vain. Even after being bitten, Budden had been insistent that his taipan be sent to the Commonwealth Serum Laboratories at the National Museum in Melbourne, where work on producing a taipan antivenom might finally start.

The taipan travelled south in a nailed-down box on which was written DANGEROUS REPTILE—DO NOT OFF LOAD. It was flown to Brisbane in the tail compartment of a T.A.A. flight, and then transferred to the Melbourne connection the same night. On the Monday, Melbourne snake man David Fleay received a telephone call asking whether he would be prepared to milk the Budden taipan. Fleay gave the request considerable thought overnight before agreeing to it the next morning. Later that day, he fought his way through the scrum of reporters and photographers at the museum. By now, the milking of the taipan had assumed an epic significance. It was being seen as the triumphant posthumous vindication of Budden's actions, the story's final episode.

Fleay described Budden's taipan as "six foot five inches long, in the finest possible condition, its orange-red eyes glittering, the white upper lip emphasizing the mouth." He and his team expressed the viscous yellow venom from the fangs through a rubber flange into a jar. They had successfully milked the snake, collecting enough venom on which to begin research. Afterwards, the team unanimously agreed that they had just "dealt with the most savage, tough and insistent snake" in their considerable, combined experience.

The story might have ended there, but the taipan had transfixed the Cairns community and the snake took up extended residence on the *Post*'s front page. On July 31, 1950, veteran snake catcher John Cann publicly declared that he wanted funds to go on a taipan-catching hunt.

"I have been catching snakes for 40 years but they still bite me," the *Post* quoted Cann. "I have been bitten 400 times by every type of Australian snake—except the taipan."

On August 3, it was reported that the venom taken from the

snake at Melbourne "in two 'milkings' was sufficient to kill 150 men." "Taipan's Deadly Record," ran the headline on August 4: "Australia's Worst Snake, Bites Like a Dog." The article claimed a 99 percent mortality rate for victims of taipan bite. It told of a lonely settler on the edge of Arnhem Land in the Northern Territory who was found lying dead on his bunk. It was evident, the article claimed, that he had "died within a few minutes of having been bitten." The settler, it seemed, had been bitten on his toe while sitting at his table. "He had tied a rough ligature around his leg, but when he felt the poison taking effect had scrawled a rough will in charcoal on the table with the words 'Bitten by a King Brown Snake.' "

Taipans and king browns are not easily distinguished. The dramatic death of the lonely settler, which sounds like an oft-repeated piece of reptilian apocrypha, had always been attributed to the king brown—until some worse species came along to oust it. It was the rapidity of the settler's death in the original story which suggested that the culprit must in fact have been a taipan. That Budden had taken over twenty-four hours to die was conveniently ignored; the taipan was in a different league to the king brown.

Suddenly, everybody in Cairns was seeing taipans. On August 4, a taipan was reported killed "by an unknown cyclist" in the grass off Aplin Street. Three days later, it was reported that three taipans had been killed in the suburb of Freshwater. Then, on August 10, a second live taipan was reported caught by council workers in a gully near the Cairns township of Stratford. One of the workers, Mr. Freeman, told the *Post* he had "killed plenty of other snakes in the past and I am sure now that some of them were taipans. We used to call them king browns." Another worker called Connolly, who had helped capture the snake, nonchalantly added, "We tied the bag up with string. There was nothing to it." At the approach to the depression, the workmen erected a sign reading "Taipan Gully."

Sightings were one thing; a second live taipan was quite

another. "Taipan Snake To Go To Melbourne," trumpeted the next day's edition of the *Post*. "Will Travel By Air. Is Still Very Lively." The capture of a second live taipan had given the story renewed impetus—and all kinds of procreational possibilities. There was breathless talk of the two taipans getting together in Melbourne. "Poison research would be simplified," explained the *Post*, "if the snakes could be bred in captivity." "Taipan Drinks Water," the paper reported on August 14. "Eats No Food. Public Showing Soon." "While Melbourne's taipan No. 1 has been drinking a daily fill of water at the museum," the paper continued breathlessly, "preparations are being made for a public showing of taipan No. 2 at the zoo later this week." Expectations, however, were cruelly dashed as the headline of the August 17 edition—"Second Taipan Snake. Found Dead in Bag"—somewhat baldly explained. "Dozens of people were disappointed at the Melbourne Zoo today," the paper continued, "as a special section had been set aside to display the snake."

A few months later, as winter descended on Melbourne, Kevin Budden's taipan also died.

David Williams turned the car round and drove me down to the nearby city cemetery. "Prime taipan habitat," said David, indicating the scrub that lined the cemetery's edge as he led me to Kevin Budden's grave. For a moment, Williams stood respectfully before the grave of a fellow snake man, then shook my hand and walked away. I kneeled to read the plaque which a friend and snake-collecting colleague of Budden's had fondly erected. "DULCE ET DECORUM EST PRO SCIENTIA MORI; He gave his life for Queenslanders; let us not forget him," it read, then embarked upon a colourfully poetic epitaph which blithely disregarded grammatical exactitude. "He laid the world away," it read, "poisoned was his rich red blood of youth, gave up his years that men call age and those that might have been...It is better to deserve honour than to possess it." In herpetology's war dead, it seemed, Budden was the Somme, Passchendaele and Gallipoli rolled into one. They would remember.

Talking of which, the sun was going down. A sweating orange orb drizzled into the clouds just above the bark-shredded gum trees. I walked back towards town through the large cemetery, with its low, rusted iron railings surrounding ornate family plots of Catholic emigrants from distant Italy and Ireland. *Prime taipan habitat*, Williams had said, so I stamped the ground with my every step to warn them off. Even so, I sensed them all around me. This was October, their spring, when they were said to be at their most aggressive. Whenever I looked around me, I half-expected rust-brown snakes to be arrowing across nearby graves towards me. Not for nothing, I thought, did Cairns have a Taipan car rentals, a Thai-Pan restaurant and an Aussie Rules Football team called the Cobras. Even the movie showing at the Ace Cinema in the Cairns suburb of Freshwater at the time of Budden's death had been *The Snake Pit*, starring Olivia de Havilland. It was by all accounts a snaky kind of place. As I quickened my steps into the gloaming, the sound of traffic was drowned in the rising chorus of insects. And as the sun disappeared, the colours of evening—browns, leads and minerals—leached back into the ground.

Then it happened. Somewhere beyond the limits of my vision, a terrible hissing noise froze my blood. I imagined the enraged snake rise at my back to fling itself forward, its whole weight concentrated solely upon driving its fangs deep into my flesh, and I broke into a wild, uncontrolled run. I did not slow until I reached the gates of the cemetery and the main road. It was only then that my mind belatedly registered what I had seen in the corner of my eye. I had alarmed a seagull. It had turned upon me, its wings puffed up and its beak agape, to defend its nest among the headstones at any cost. From the haven of the pavement, I looked back at the cemetery. I was sweating profusely, my pulse was racing and my eyes had acquired a wild stare. From cars, people were regarding me with bewilderment and suspicion. Some of them even wound up their windows. So, in an attempt to regain some dignity and despite my thoroughly unsporty clothes, I checked my watch as if timing my progress and moved off down the street with what I hoped passed for a casual exerciser's trot.

7. AFRICA II

The man who can caress a snake can do anything.

—Isak Dinesen, *Out of Africa*

JUNE 1996

The train approached Mombasa at dawn. Barefooted children arose from the soft-lit landscape like clouds of flies to run wide-eyed and screaming with the passing carriages, their feet slipping on the black earth where the night rain had slicked it. The woodsmoke that billowed heavy with steam from village fires had a sodden, vegetable smell. At Mariakani, the stationmaster shooed chickens off the line to stand proud at our passing. He wore a pith helmet and held a furled green flag tight against his thigh. His white, high-collared uniform was immaculately starched. Only an old black stain from a top-pocket biro let him down.

I caught the first bus north along the coast. Through the morning drizzle, I looked out on suburb billboards that trumpeted "Construction materials in fibre cement," then on sisal plantations and dirt roads disappearing among the cashew trees, while the ocean remained unseen, always a scrubby hill away. I was heading for Watamu, a resort village that I knew to have a snake park. But at Kilifi, where an impressively modern bridge carried us across a turquoise creek, I remembered how yesterday's attendant at the Nairobi snake park had mentioned the black mamba—bite survivor here.

According to William Fitzgerald, a pioneer explorer of the Kenya coast, the Kilifi area was "noted for its many poisonous snakes." His Kilifi camp was surrounded by high grass; "it was," Fitzgerald reported, "rather risky going about." He had been dispatched by the British East Africa Company in 1891 to assess the company's first few plantations—mostly cotton, jute and to-bacco—and to report on the region's potential for further agri-cultural expansion.

A surviving photograph has Fitzgerald wearing a monocle, a droopy moustache and a battered fedora; there is something of the dandy about him. But he also wears the regulation khaki and leather gaiters, holds a bolt-action rifle and is sitting on a bench that is draped in what appears to be an impressive python skin, doubtless his own "bag." The photo is artfully framed, with a couple of water bottles lying at his feet to em-phasize the stripped-down needs of this redoubtable man of Empire, an adventuring agricultural consultant.

For all this, though, Fitzgerald was not immune to troubles, particularly from snakes. Some of his experiences with the snakes of the Kenya coast left him, in his own words, "in a ghastly funk." Fitzgerald spent Christmas Day of 1891 in camp on the Galana River, in the Malindi district to the north of Kilifi.

This half-way camp [Fitzgerald wrote], was an awful place for deadly snakes...In the morning, just after I had got up, I saw one going under the mats of my tent, and on lifting them up, I found a hole under the head of my bed. Shortly after-wards, my servant found another coiled up amongst my boots; it escaped before he could kill it. In the afternoon, whilst I was sitting reading in the store shed which I found cooler than my tent, with a cup of tea on the table by my side, I suddenly heard my spoon jingle, and looking up hastily, I saw a large green "mamba" with its nasty flat head over the tea cup, its body curled round my chair and table. I gave an involuntary jump, the snake turned rapidly towards me with a hiss; I remained stock still; then to my intense relief, it slowly uncoiled and glided away. I am not ashamed to say I was...quite unnerved for the moment. Half an hour afterwards, I heard a rustling at my feet, and looking down I saw a yellow-spotted brown snake gliding between my legs . . .

We came to a halt in Kilifi's market square. Villagers squatted by neat pyramids of little snouted mangoes, thick stalks laden with bananas, baskets of eggs and piles of mangrove charcoal. I escaped from the hubbub and came to a nearby ironmonger's, whose claim to be the "Main Urgent for Crown Paints" spoke of respectability; the sort of place I could safely leave my bag. I could also ask the ironmonger about snakes.

"No," he said, mistaking my general query for a bizarre stock request, as if his were some kind of pet shop. "But we have gardening tools."

"I mean," I continued, "what do you know about them?"

"A snake park there was down by the river," the ironmonger told me, an arm winding across the counter. "But it is sadly clos-ed now."

I followed his directions to the bridge, down potholed streets lined with shacks roofed with corrugated iron or thatched with *makuti* palm, past lumber yards and monkeys in the mango trees. A man on a bicycle, the Black Mamba as they would have it in Nairobi, gradually overhauled me, but he was going at anything but a black mamba's pace. He was pedalling far more furiously than his modest speed seemed to warrant. Evidently, the sole gear that still operated was a high one.

By the time I reached the bridge, the sun had broken through. I looked down the creek to the ocean, where the waves ruffled and grew tall. Inland, the calm lagoon was hemmed in by high-sided, wooded bluffs where the beach houses and yacht clubs of wealthy Europeans were perched. A young man with bright eyes and a freshly starched blue shirt approached me.

"My name is Jacob," he told me. "I hold a diploma in business admin. I speak English, German and Spanish. I am a tour guide." But like Kilifi itself, the multi-talented Jacob had been short of work ever since the bridge was opened in 1991. Jacob explained how the ferry crossing had been a popular feature of this coast. Crossing had made tourists feel like true travellers, as if the ferry heralded adventure on the far bank. It had also stranded them in Kilifi at the end of the traffic lines with little to do but be shown such sights as the town's south-side could offer.

"Now, they drive straight through. To Kilifi, they represent no business opportunities," said Jacob, downsized by progress. "So why are you in town?"

I told him I was looking for a black mamba—bite survivor.

"Ah," replied Jacob with a courteously broad-minded equanimity. "*Bwana Nyoka*. The snake man. There was such a man. Come with me."

He led me off the southern side of the bridge, and down a track to the creek's edge at Mnarani, a place of sudden silence mantled in the dark greens of a patchy forest canopy. Mnarani had taken abandonment hard. We were standing on what had once been the coastal highway between Mombasa and Malindi, where the ferry traffic had regularly lined up for hours. Then they built the bridge, which did in the ferries and a whole host of dependent businesses. The new road to the foot of the bridge had been diverted a few hundred feet away. Mnarani was bypassed and had become a Kilifi backwater taunted by reminders of its former bustle. Doors slammed in roadside bars where only the wind wound. The loose goat that cropped at grass along the side of the road to the old ferry landing had long since stopped pausing in its grazing to check for traffic and the dangers it represented. And the once-bright tops of the soft drink bottles, embedded in the tarmac in busier times, had long since tarnished and turned to rust.

"This was where the Bwana's snake park was," said Jacob. We

were standing below Mnarani's old Arabic ruins, the remnants of a mosque among a fallen litter of inscribed stones in a shadowy cavern of trees. Jacob was pointing to a small plot behind a fence of barbed wire that had rusted to a brittle filigree where somebody called Khan—the painted board on the tiny shed revealed—now grew bananas. Nothing remained of the snake park; no evocative abandoned display cases, no labels lying in the dirt that indicated the cases' former contents and their bracketed taxonomic names; puff adder (*Bitis arietans*), black mamba (*Dendroaspis polylepis*) or black-necked spitting cobra (*Naja nigricollis*).

I asked Jacob what he remembered of the European snake man.

"I never knew his name," said Jacob. "We only knew him as Bwana Nyoka, the man who survived a bite from a *tsasapala*, a black mamba. But that was a long time ago. When I was little. Of course, we were afraid of him."

I asked why.

"Snake men have a power over snakes. A power over snakes, that's a strong power. Like the power of our *mchowis*."

"Your what, Jacob?"

"Our mchowis. Witch doctors. The ones that send snakes to deal with people."

"To *deal* with people?"

"Yeah. You know, if somebody's done something wrong, some illegal business practice perhaps, then the community will ask the local mchowi to organize a snake."

"To deal with that somebody?"

Jacob nodded.

"Do you know the local mchowis?"

"They keep to themselves. But I could find out."

"Jacob," I said, already planning a return to Kilifi. "Do you want some work?"

Jacob accepted my commission gratefully and disappeared to pursue it. I walked down to the old ferry landing, where a few small fishing smacks were moored. In one, a man sat, picking attentively at his toe. When I asked him whether he would ferry me across, he looked at me in bewilderment, wondering whether the fool mzungu could possibly be unaware of the toll-free bridge at his back. Quickly, he motioned me towards the boat. To protect his prospective fare, he got to his feet and puffed himself up, concealingly bridge-shape, as he did so. Light ripples patterned the water as he rowed me across the creek. The shy smile on the fisherman's face grew broad; then he began to have trouble controlling himself. Eventually, he collapsed in a puddle of nasal snorts and, aware that I might feel I deserved an explanation, pointed seawards.

"Bridge! Bridge!" he yelled, and slapped his knees, almost losing an oar into our gentle slipstream. How could I not have known of it?

"I'm aware of the bridge," I muttered, irritated. "Old times' sake," I added. But the explanation was hopeless. With a final shove of the oars, the fisherman propelled the boat in to the shallows and, his commission completed, toppled backwards off his seat into the bottom of the boat, howling with laughter.

FEBRUARY 1997

"A man called Khan," murmured Pete, puffing on his pipe in Banchory's Burnet Arms and peering into his coffee. "Growing bananas, eh?" He chuckled. "When I bought that land, in early '64, there was an old tea house on the site."

He had just completed five years in Tanzania's National Parks and Game department, and was scratching around for a new living. He had arrived on the coast to open his snake park in December 1963. All along the road from Mombasa, he had watched people slinging the lianas of black, red and green bunting, the colours of independent Kenya, between their windows for the independence celebrations, while government workers erected wooden platforms for the speeches of their leaders.

But from his girlfriend Jan's house at Takaungu village on the next creek south of Kilifi, Pete soon found himself thinking less about Independence and more about the lines he had experienced at the ferry crossing on trips to Kilifi. He sensed what Jacob would have called a business opportunity.

"To me," said Pete, looking into space and pointing the stem of his pipe in the direction of the Burnet Arms's fruit machine, "it meant a captive audience—locals and tourists alike—just waiting to be entertained."

The Kilifi ferry landing was where Pete would build his snake park.

By February 1964, he had bought the land and construction was under way. The architecture was inspired by Pete's memory of his school classrooms at Naivasha; a makuti thatch above frames of podocarp wood with walls of split sisal poles nailed vertically inside and out. Pete omitted the insulating mud with which the walls at high-altitude Naivasha had been filled; Kilifi needed all the breeze it could get. Within weeks, a twenty-foot square building, with lean-tos running along several of the external walls, had been completed; this was where the snakes would be displayed. A small office was also built for the attendant. Its door frame was constructed from mangrove poles, with a corrugated iron panel. Wooden brandy and whisky cases were requisitioned from Mombasa shopkeepers. Pete turned them into display cases. He gave them glass fronts and, a feature he was particularly proud of, sliding partitions down the middle so he could safely clean each side of the case without going to the trouble of removing the snake. He filled the cases with a layer of shale collected off the beach which he then covered with dry mango leaves.

Kilifi children, drawn to the activity down at the old tea house, were among the first to discover the news. In the days before Pete opened for business, gaggles of screaming children could be seen running out of the gates to alert their friends: Kilifi was to have a snake park. Local foreigners, anxious about their revised status in newly independent Kenya, were ripe with supportively shrewd suggestions.

"Education, education, education," Nurubai, the Indian shop-keeper, told Pete. "Just make sure you are getting plenty of African school children to visit," he confided. "No one is touching you if you make it educational."

By early June, the snake park was almost completed and Pete had heard all the advice he could use. Though what he could do with were a few more snakes. He had collected some of his own, but clearly needed some high-profile specimens to lend the venture an initial credibility. He took the train to Nairobi where Uncle Alan provided him with two pythons, two cobras, a hissing sand snake and a rhombic night adder. "And you'll need mambas too," said Uncle Alan. "You can't have a snake park without mambas. But you'll have no trouble finding them round Kilifi."

The Kilifi Shack, as Pete called his snake park, opened on June 21, 1964. There was no fanfare, just a man on the gate to take a few shillings and some snakes to admire. Pete took a trip into Mombasa to pay courtesy calls on tour operators Rhino Safaris, Pollmans and UTC; their clients, he suggested, might find the new snake park an interesting distraction during the ferry wait.

The Kilifi Shack flourished, attracting 200 visitors on the better days. The tour operators proved supportive; Pete delighted the tour guides at the nearby Mnarani Club by naming two of the pythons Jimmy and Suzy in their honour. Pete's first black mambas, which he caught near Kilifi in the latter months of 1964, particularly seemed to transfix the German tourists. Pete also had much to do at Takaungu, building holding pens for the breeding snakes and also for the rabbits and rats that he bred as feed for them. Pete was handling a lot of snakes, but it was not until late the next year, 1965, that he took his first bite.

"I thought it was a harmless snake—a pretty little thing called a Cape Wolf," he explained. "I pretty soon realized it was a mole viper when it bit me on the hand."

The mole viper is not the most life-threatening snake, and all Pete had felt was a brief stab, but it was his first bite and he was not inclined to delay getting himself to the hospital. He climbed into the car and drove down to the ferry, past the wooden kiosks selling cigarettes and Fanta, tea, newspapers and souvenir seashells to the front of the traffic line. Furious drivers sounded their horns in protest. He arrived at the landing-stage just as the ferry was docking, but the skipper had spotted the snake man's maneuver and waved him away with a back-handed dismissal. So Pete shoved an arm through the car window, banged the roof to get the skipper's attention and then jabbed at his arm with the fore and index fingers of his other hand.

The ferry skipper had seen enough. Immediately, he motioned Pete to board. The ferry was soon gone, propellers boiling at the water and Klaxon sounding above the enraged car horns of those who had been left behind. The skipper selected full ahead; it was not every day he had a snakebite victim on board.

JUNE 1996

"Watamu, Watamu, Watamu," yelled the driver of the Double Happiness *matatu* minibus. He had spotted the Europeans among his fare. "Beautiful beach. Lovely village. Fine hotels; we got Hemingway's and Ocean Sports—that's Open Shorts to some. And we got our snake park." The passengers rose to the driver's showmanship with a collective, theatrical shudder.

"Was that an Oooh Nooo I just heard?" joshed the driver. "Laydeez, laydeez!"

He dropped me last, past Watamu village and the string of beach hotels, way down the spit road by a sign that read "Plot 28; Simpson," and left me to shoulder my bag and walk the few hundred yards down a sandy track towards the sound of the sea. A collection of low whitewashed buildings, of no discernible style but entangled in hibiscus, bougainvillea and pink English roses, came into view around a final corner. An elderly European lady was sitting on a verandah in a *kikoi* wrap, a Walkman clamped upon her head.

Barbara Simpson threw off the headphones and smiled.

"You must be the chap about the snakes. You just missed Jimmy Ashe, our local snake man. Just left. *And* ran over poor Bumble on the way out, I might add." Bumble, a short-haired dachshund, was shivering with shock under his mistress's chair.

Barbara had bleached blonde hair, bright brown eyes and a disfigured face which almost became her in old age. She had spent most of her life in Kenya since stepping ashore with her family at Mombasa in 1923.

"I was eight at the time," she would tell me later. "We had sailed from India. And somewhere off the Seychelles, two boys about my age threw Dulcie, my dolly, overboard." Almost seventy-five years later, you could still hear the affront in her voice. "There was a kind lady who noticed me crying and actually made me another Dulcie. Lord knows where she found the materials on a packet steamer in the middle of the Indian Ocean. Still, I'll always remember her."

She had lived at Plot 28 since the 1950s, and had watched a bustling resort emerge from Kenya's littoral wilderness. Fancy hotels and shops had been built. Tourists had started to arrive from Europe. Following the lead of Ernest Hemingway, big game fishermen had pitched up from Zimbabwe and Uganda. Africans from the interior had come in search of work. They built curio kiosks and *shambas* or smallholdings on the unwanted plots along the land side of the road, where chickens pecked at the dirt and shabby stands of maize grew. From their mud and makuti huts, they looked across the tarmac that separated them from the expensive hotels with their uninterrupted views of the ocean.

From Barbara's verandah a sandy path led down to the beach. Beyond the casuarina trees lay a coral sea in which a small dinghy bobbed at its mooring. Watamu's famous coral reef lay just offshore. Ferdinand, who worked for Barbara, took guests snorkelling in the mornings.

"Client relations, that's Ferdinand," said Barbara, clarifying the roles of her members of staff. "Then there are the Jonathans. There's Jonathan B., chauffeuring, plantlife and wildlife; Jonathan C., compost and carpentry; and Jonathan K., cooking and accounts."

So could Jonathan B., I wondered, help me out with snakes?

"Oh, he's an African," said Barbara. "They detest the things. Jimmy's your man. His collection is wicked. I do beg your pardon, I mean splendid. I've had a couple of students out from Manchester studying mangrove conservation. I seem to have picked up some of their expressions. Can't say I'm so taken with their music though." She gestured at the Walkman. On the cassette box that lay on the table, the words "Wicked Techno" had been scrawled.

In the afternoon, while Barbara slept, her head pulsing with unfamiliar musical rhythms, Jonathan B. ran me up to Jimmy Ashe's place in Barbara's little white pick-up. We passed hotels and pastel-painted African churches. Roadside *dukas*, the wooden kiosks that sold shaving cream and cigarettes, advertised themselves, in white paint, as "Tesco" and "Marks & Spencer, Watamu Branch."

On Watamu's north-side, walled plots contained half-finished villas in Spanish hacienda and faux-Tuscan styles, all pink bell-towers and loggias, terracotta roofs, decorative ceramic tiles and Gaudiesque flourishes against a background of ocean and

palm trees. Each property, even the uncompleted ones, already
boasted a post-box which was as neatly European as the name
stencilled upon it.

Jimmy Ashe's snake park stood in the midst of this displaced
suburban idyll, oblivious to the changes around it. Jonathan B.
dropped me at the entrance. I walked into the driveway which
disappeared down a tunnel of greenery into a cool, shaded
world where the sun appeared in brief dapples through the
thick canopy. No foliage, it seemed, had ever been cut on the
Ashe plot. No whining Weed Whacker had ever broken the silence of the afternoon. Jimmy lived deep in a forest. And one
day, you sensed, it might overrun him and his household completely.

The snakes were housed in glass-fronted, duck-egg-blue display cases that sat on outdoor shelves and trestles, flanking the
latter stretches of the driveway. They made for an intimidating
approach. Jimmy was sitting on his verandah. Like Barbara
Simpson, he wore nothing but a faded kikoi wrap. He was bald,
with a patriarch's grey beard and piercing eyes. It was a strong
face which must once have been winningly handsome. He reminded me of the elderly Hemingway, who had once chased
marlin off this coast and had subsequently had a local hotel
named after him.

Jimmy patted me to a seat. He was convalescing after the re-
moval of a cancerous lung. He was short of breath and much of
the work at the snake park—showing tourists around, milking
the mambas of their venom, repairing the snake cases—was
now done by others. It was his mind, he told me, that had really
taken a hammering from the operation. Names eluded Jimmy
these days, dates flew away, sentences petered out in a frus-
trated series of half utterances.

He knew his snakes though. He had just about every Kenyan
species on display at his snake park, and a milking herd of
about sixty green mambas with which he supplied antivenom
manufacturers. Greens seemed to love the coast; the place was

alive with them. Blacks less so; they became more numerous the moment you ventured inland. Couldn't respect blacks enough, though, whenever he came across them.

"We had a fair number of black mambas on show up at Nairobi," Jimmy told me. He had once been curator of the Nairobi snake park. "Used to go into the cage containing cobras, seventeen of them, to muck out with a broom and hose. A room measuring fifteen foot square. Would never think of doing that with half the number of blacks. Once sent ten blacks to the U.S.A.," he explained, "and within a week one managed to kill the keeper."

Jimmy had fond memories of his Nairobi days. The snake park had attracted many famous visitors. There was, there was . . .

"Oh, you know," he ventured. But I didn't, and he clapped his head in frustration. "That flying chap," he said. "You know."

"Richard Branson," I suggested. "Douglas Bader? Freddy Laker...?"

"LINDBERGH!" boomed Jimmy, slapping his knee. "Met him." He stared into space for a moment. "And the Aga Khan. And that pop singer. Beatle, I believe."

"Paul McCartney," I ventured.

"Right first time," said Jimmy. "Thought he was some soldier on leave from Aden, I should say." He shrugged. "Even Beatles seem to find snakes fascinating. Which is why you've got snake parks all along this coast; Easterbrook's at Malindi, the one at, um, um, Bamburi just north of Mombasa, the ones belonging to various African chaps, and that chap, that chap...oh, you know, it's clean gone, the chap that got clobbered by a black mamba down at Kilifi. Terrible time he had of it, days in the hospital, but survived somehow. Damned fine snake man he was too. Dead, he is. Or is he? Moved away at any rate. Bartley? Was it Bartley?"

Jimmy Ashe couldn't remember Pete Bramwell's name, but in time I would come to see how much the two had in common.

East Africa's European snake men shared many things, not least an affection for whiskery, Victorian facial hair. They had both worked in wildlife. They both disdained culinary expertise, avoided vegetables at any cost and clearly didn't dress to impress. They could look pretty shabby if they put a mind to it, could Jimmy and Pete. They were mavericks that often seemed to prefer the company of snakes to humans, despite the bites they had taken. They were old-style colonials, often irascible, reactionary and eccentric, but their attachment to Africa was almost visionary and their commitment to snakes absolute.

They were a type, a type closely modelled on East Africa's original white snake man, a self-professed Englishman of Greek parentage who had made snakes his life 600 miles to the south in former Tanganyika a generation earlier. The legendary C. J. P. Ionides owned a puff adder called Popkiss and ate gaboon viper for his Christmas dinner. He was regularly seen wandering around the African bush wearing snorkelling goggles. The fact that there was good reason for this—they protected him from the spitting cobras he was pursuing—did nothing to dilute his singular appearance.

These men were undoubted eccentrics, but what most distinguished them were their rigorously practical minds. Their greatest pleasure in life was in meeting the needs of their snakes—encouraging them to eat and breed in captivity, treating them with pyrethrum powder for that scourge of snakes, crows' mites, capturing and transporting snakes without doing them harm, milking them efficiently, customizing boxes and cases for them, and making specialized adaptations to catching tongs and snake hooks. Much of what they knew they had learned at first hand. They saw themselves as reptile vets, breeders and general herpetologists. And in this modest field they excelled. They considered pioneering advances in snake husbandry as the great achievements of their lives.

To the superstitious Giriama of the Kenya coast, however, the white snake men were something else entirely. The Giriama saw them less as herpetological crackpots than blessed with special powers. They respected, even feared them, for only the select few handled snakes. The rest simply avoided snakes. They feared what they could do. They might leave offerings out—a bowl of milk, perhaps—to placate passing snakes or to nourish ancestors incarnated as snakes. To kill them was widely regarded as a reckless act that invited vengeance. The heads of the few snakes that were killed were often valued as ingredients in clandestine dawa potions. Such nostrums were often intended to harm, and many Africans buried snakes' heads to prevent them falling into the hands of mischief-making mchowis while unscrupulous others turned a profit by selling the heads to them. William Fitzgerald, who was forever killing snakes and often ordered them skinned, noticed how the head was frequently absent when the skin was presented to him.

Even those Africans who handled snakes, the mchowis and other occult initiates, found the experience traumatizing. According to William Fitzgerald, one of his entourage, an African snake man called Ambari, suffered "a trembling fit, and if he could, would get drunk on coconut toddy" whenever he handled snakes.

By contrast, the power that the white snake men possessed over snakes seemed boundless. They set up shop all along the Kenyan coast, and made snakes their business with a thoroughly professional calm. They made it known that they wished to be notified of any snake sightings with a view to capturing snakes and taking them home. They would even offer a few shillings for information leading to successful captures. They actually seemed to like snakes. And it wasn't just the snake men: plenty of other whites proved interested amateurs. When Fitzgerald's toddy-drinking Ambari turned up one day "dragging a puff adder on a piece of string," it was entirely characteristic of the European attitude that Fitzgerald, despite the "ghastly funk" that earlier sightings had left him in, should take the snake from Ambari, put it in a box and have it dispatched to the reptile house at the London Zoo. It was, it seemed, scientific curiosity putting superstition to flight.

It was in a similar spirit that the European snake men studied their snakes' scale arrangements and their fang apparatus, weighed them and wrote papers about them for obscure herpetological journals. And they measured their speed. Jimmy Ashe and Ionides had once measured the speed of a black mamba. The turn of pace of the world's fastest snake has long been liable to exaggeration, and establishing the black mamba's top speed on a scientific basis became the singular ambition of a number of herpetologists.

"We'd had enough," said Jimmy, "of all these stories of mambas outrunning horses and bringing down their Boer riders. And, frankly, that afternoon we'd also had enough of each other, old Iodine and I; he liked being called Iodine did, er, Ionides. I mean, I said fifteen miles an hour, maybe more; he said less than ten. Impasse. To stop us coming to blows, Iodine then suggested we put them through their paces. I had a couple in bags at the house, you see."

It was 1959, when Jimmy was living at Athi River, about thirty miles south-east of Nairobi. With the help of a few whiskies, the afternoon had turned lively. Now, as talk turned to the speed of the black mamba, Jimmy's guest was coming over cantankerous. Between them, the two had caught thousands of mambas, and each had caught quite enough of the combined total, and drunk quite enough whisky, to scorn the other's opinion. With the help of a tennis line marker, the warring experts drew two concentric circles on Jimmy's lawn, with radii of twenty-five and fifty feet, and placed a bagged black mamba at the bull's eye. Jimmy collected snake hooks, pens and paper, and raided the linen cupboard for two white pillowcases. To lend the experiment credibility, he even managed to lay his hands on two stopwatches.

When they were ready, Jimmy released the drawstring on the snake bag and stood back. The released mamba seemed to move like an arrow. To push it to its limit, Ashe and Ionides chased it, waving their pillow cases at it with all the energy of sweethearts watching their loved ones go to war. As the mamba reached the inner circle, the men started their stopwatches. When it crossed the second circle, they stopped their watches, immobilized the mamba with a snake hook and returned it to its bag.

For the sake of accuracy, the second snake was put through its paces while the first mamba rested. Each snake was timed on three occasions in the course of the afternoon. The men then freshened their whiskies and sat down over their sums. Independently, they arrived at speeds of between seven and eight miles an hour. Ionides had been proved right.

It was a memorable cameo, but it was not the first recorded attempt to confirm the true speed of this fabled snake. Richard Meinertzhagen was a soldier who served in Kenya in the early 1900s, achieving notoriety for his brutal suppression of the natives and for claims, posthumously established as fraudulent, regarding the provenance of specimens in his extensive stuffed-bird collection. Meinertzhagen's remarkable Kenya diary faithfully records the number of game he regularly observed from the window of the Nairobi train. And judging by the numbers he gives—1,207 zebra, 3,715 wildebeest—these were clearly head counts, censuses rather than estimates. For the sake of his fellow passengers, one can only hope that Meinertzhagen was not in the habit of writing out loud.

It was no surprise then that establishing the true speed of the black mamba should have appealed to Meinertzhagen, whose interests as naturalist and statistician were equally obsessive. The opportunity to measure the snake's speed presented itself to him one afternoon in April 1906 while patrolling the plains of the Northern Serengeti, when a large black mamba appeared ahead. Meinertzhagen was not equipped to immobilize the snake, which made a standing start out of the question. Initially, however, the major problem seemed to be that the snake was not inclined to move. Unimpressed, it raised its head clear of the ground and turned to face the men. If it would not flee of its own accord, Meinertzhagen figured, he must instead provoke it into pursuit. He would cause the snake to chase one of his *askaris*, or soldiers, and time it that way. (It is to be hoped he had already convinced himself that his man could outrun the mamba; being on the Meinertzhagen staff was not said to be much fun.)

Meinertzhagen's patrol incessantly baited the snake with clods of earth. Eventually, the infuriated snake turned on an unfortunate askari who headed off across the plain in a straight line as instructed. The triumphant Meinertzhagen timed the snake's formidable progress until the pursued askari tripped and fell after what Meinertzhagen later measured as forty-seven yards. For accuracy's sake, Meinertzhagen would have preferred to have timed the mamba over a greater distance, but he reluctantly acknowledged that it urgently needed shooting if the askari was to escape with his life. Like Ashe and Ionides half a century later, Meinertzhagen gave the black mamba a top speed of seven miles per hour.

Where African superstition had long seen mambas as spirits, European science tended to regard them as subjects of herpetological, neurological and toxicological study. Most remarkable was one Dr. F. Eigenberger, whose clinical interest in the action of poisons led him to spend much of the 1920s injecting himself with snake venoms, pen and paper poised. Having experienced for himself the effects of rattlesnake venom ("painful swelling...local symptoms of a very painful character"), Eigenberger prepared to inject himself with venom from the West African green mamba (*Dendroaspis viridis*), not quite as potent as that of *Dendroaspis polylepis* but a deadly venom nevertheless.

Eigenberger took a single drop of the green mamba venom, diluted it in ten times the quantity of salt solution and injected a small amount of the dilution into his skin. His extensive notes on this experiment, which appeared in the *Bulletin of the Antivenin Institute of America,* make compelling reading; "a burning sensation…slight local itching…inflammatory swelling…a very distinct over-irritability of all sensory nerves." "The vibration and motor noise of my car," he wrote, "appeared so loud and annoying that I thought all four tires were flat and it made me stop and look before I realized the reason." One wonders, of course, why Eigenberger should be driving a car so soon after injecting himself with deadly snake venom, but then much of Eigenberger's wider behaviour causes one to wonder. He wrote of a "slightly intoxicated feeling, with some fainty weakness," then felt "weak and deathly sick," and experienced "a numb feeling around the lips, chin, and in the tip of the tongue, spreading rapidly over the entire face and down the throat." The eyes began hurting. There was humming in the ears. Numbness invaded the fingers and toes. The pain moved into the chest. Talking and swallowing became difficult. Breathing became increasingly painful. The forearm and hand on the affected limb were intensely swollen. Two hours after the injection, the numbness over the body was turning to a general pain. "This proved the most curious and interesting effect," wrote Eigenberger. "Five or six hours later," he continued, "the whole body, wherever touched, was extremely painful." He suffered flu symptoms during the night but was on the mend by morning.

For all Eigenberger's sufferings, it was the mambas I felt sorry for. With the coming of the Europeans, they suffered a rude awakening where they had previously had it made. Africans noticed them sunning themselves on graves, tipped figurative caps at long-departed ancestors and kept their distance out of fear and respect. Black mambas needed only to rear up, throw their hoods and you couldn't see the Africans for dust. For years, places where black mambas had been sighted would become strict no-go areas for humans.

Then the white snake man came, all bags, snake hooks, inquisitiveness and lucrative contracts to supply antivenom manufacturers. And, like the poor cartoon ghost, the mamba found it had lost its tried and tested spooking ability overnight. Rather than avoiding them, these men were actively seeking them out, offering money for information on their where-abouts, capturing them, milking them of their poison and sub-jecting them to all kinds of ignominies in the name of science. What good were respectful write-ups in these men's memoirs compared to the venerated quiet life they used to enjoy from the Africans? For the black mambas, the coming of the colo-nials had meant quite a falling off.

Darkness had fallen over Jimmy's forest, and he insisted on dropping me back at Barbara's.

"Until recently," said Jimmy, "you wouldn't have seen Africans on this road after dark for fear of the *shaitani* or spirits. Tourism and tarmac have begun changing all that. But these are still superstitious places." Jimmy smiled at himself, remembering something.

"A few years ago," he explained, "they tried clamping down on the mchowis in the area. Rational progress and all that, don't you know. Every mchowi had to hold a license. They told me I needed one. I explained that I was not a witch doctor. And the officials looked at me. 'Of course, you're a mchowi, Bwana Nyoka. You handle snakes, don't you?' "

Jimmy drew up at the turn-off for Barbara's drive.

"Can I ask you to walk it?" he said. "I can't face the prospect of running over Bumble again."

He turned the car in the road, the lights nosing deep into the bush, and was gone, the rattle of his engine merging into the sounds of the night. I was reminded of the last time I'd walked in the night through snake territory—outside Carl Porter's church in Georgia—when a crucial difference struck me: I did not have my torch with me. And there was no moon. I cursed myself for a fool, and thought of the track, several hundred yards long, that separated me from Barbara's house. I stood in the darkness, feeling like an idiot and thinking of the path ahead. And the smug words of Elijah Bingham swam into my head, mocking my predicament: "Often at night, especially in the early rainy season, we have almost trodden on little snakes in our path...there are little snakes in your path and mine daily, ready to bite, and some bites may be fatal...We have carried a torch or a lantern, and dazzled the snake with the light till someone killed it."

But there was no light, at least no more than my match which the breeze soon snuffed out. I lit a cigarette and held it low in front of me, but it hardly threw enough light to illuminate my quivering hand, let alone the way ahead. In desperation, I pulled some pages from my notebook, twisted them into a plait and set light to it. No flaming firebrand resulted.

Eventually, I accepted the inevitable, pulled deep on the cigarette and stepped forward into the darkness. I trod softly on the sand, and then remembered that puff adders were notorious for not moving out of the way, so I trod harder in an attempt to disable whatever I was about to tread on. I progressed like an infantryman, stamping my way along the sandy track while the sounds of the night—the birds, the monkeys, the insects and the distant surf—mocked me with the drift of their benevolent melody.

My feverish imagination was going haywire on me. It had been the custom of the Luo people of Western Kenya, I remembered, to tie the tails of live mambas to trees by paths where their intended victim was expected to pass. Enraged by their imprisonment, the mambas were sure to strike at anything which came within range. The Luo apparently found this an excellent means of securing fresh meat since a mamba bite would bring down an adult buffalo but would not corrupt its flesh, and, more relevantly, it was an equally excellent method of exacting hands-off vengeance. I wondered whether there were any Luo among Barbara's household. How did you recognize them? Had I been inadvertently rude about some aspect of Jonathan K.'s cooking, Jonathan C.'s carpentry—some muttered, much regretted comment about rot in the eaves, perhaps—or Jonathan B.'s driving? I hadn't laughed at Jonathan B.'s overtaking, had I, on the way up to Jimmy's?

I felt alone, fog-bound on cliffs, expecting the worst at every step. But there came no thrashing around my ankles, no sudden strike, only the lights of Barbara's house finally showing through the trees. I stretched my steps towards the light. Then, it was falling upon the track, faintly at first but with increasing strength until I could see clearly in front of me and the foliage on either side seemed to shrink back, and I had made it. Barbara was still sitting on the verandah, puzzling over the Manchester techno tape. If she had not been distracted by the music, she would have seen a man with clenched fists and staring eyes approach the bright pool of light around her verandah and shout "Yes, yes, yes!" as he pogoed triumphantly up the steps to safety.

8. AMERICA III

". . . the White Fang Society, an honor reserved for those who get bitten by a rattlesnake, and live to tell the tale."

—*New York Times*, April 16, 1966

JANUARY 1996

But let me take you back to the beginning and my first American encounter with snakes; a rental car making its way out of Atlanta Airport to join Interstate 75 heading south on a winter's morning. Vacation traffic bound for Florida sweeping down through the few final wooded folds of the Appalachians. Buicks and Chevrolets from Indiana and Ohio nosing past, their rear windows a collage of sun umbrellas, bedding, golf clubs and squashed children's faces.

But what I also noticed that morning, because I was on the lookout for such things, was how several front windows had been wound open a couple of inches. Sun-starved northerners imagining that first, reviving breath of the balmy south reaching them from America's subtropical hem, I guessed; a fanciful waft of sea salt, peaches, muscadine grapes and live oaks hung with swathes of beardy Spanish moss. Dreamers every one of them; I could smell nothing but the mist, wet and rubbery. But, then again, it was certainly milder than it might have been. And I had very good reasons for wishing it cold.

In the writings of the renowned herpetologist Laurence Klauber, who devoted many years to the study of rattlesnakes, I had recently come across the passing observation that "the rattlesnake hazard is reduced or completely eliminated in cold weather."

Klauber, who claimed to have offered this information "as a matter of practical interest to hunters and fishermen," could have had no idea that it would also make mighty welcome reading to a snake softie who was not enjoying the prospect of facing up to snakes. For weeks, I'd been appreciating the ugly truth that the only snakes I'd actively approached had been from the sensible side of the glass on the front of televisions. At that time, I hadn't yet graduated to the reptile house at London Zoo. Steeling myself to go looking for them, as I must surely do, was not proving a great success. And then I read Klauber's wonderful sentence, and the world resolved itself into a little ball that rolled at my bidding. Gerard Krefft's 1869 volume *The Snakes of Australia* offered further reassurance. It described boys laying out stones on the north side of Sydney to attract venomous snakes in the winter months; "snake-hunting has been a pastime here with school-boys for years," wrote Krefft as if describing the joys of butterfly collecting. So snakes were hopeless in the cold? January, then, I thought, making a note in my diary with such enthusiasm that the broken pencil lead littered the page.

This, then, was what made the open windows of those Buicks and Chevys so alarming. On this, my first snake outing, cold was the element I'd banked on to keep me from hysteria. Cold promised me controlled exposure to snakes, which was exactly the kind of exposure I had in mind. These vacationers could have all the warmth they liked down in Tampa and Miami, just so long as the high temperatures didn't creep north into South Georgia, where I was headed. Warmth I could truly do without. I hadn't come for the usual Southern ingredients, the pitchers of lemonade on hot nights and the whine of mosquitoes, whirring ceiling fans and bugs crashing into screen doors. I'd come for rattlesnakes so cold they couldn't touch me.

The Georgia countryside was flatter now, unfurling towards

the Gulf of Mexico. Woodland and soybean fields abutted the interstate and stretched, skin-tight, into a cowgum-coloured distance. On the horizon was an occasional wintry copse where spindly trees had clubbed together to daub a high, island bluff. I tuned the rental car's radio to the local stations, peopling this landscape with the pitches, promises and plain counsel of small-town opportunists and demagogues. Near Macon, an ad asked: "Is there a cat food that helps maintain cats' urinary tract health?" The delivery was smugly reassuring; *trust us*, it said, *to know just what catchment-area cat owners get to worrying over.* On one station's "pre-owned notice board," callers traded Rottweiler puppies, oak firewood, small bales of peanut hay and a '71 Volkswagen Beetle (for parts). "I have a lot of clothing," a diffident man informed listeners, "if anyone's interested." When an interested caller asked for the size of the clothes in question, they got the diffident man back on the line. "You jus' hang on there," he told the presenter. "Jean," you could hear him calling cheerily in the background, his voice echoing up a lonely staircase, "what size *was* Papa?"

"The Word of the Lord is being changed," ranted a country preacher. "You got preachers think there's nothing wrong with homosexual marriages. You got girls wearing short shorts to church, with no more respect for the House of the Lord than the man in the moon—and there ain't no man in the moon. And out in Hawaii," he hollered, as if this were the final depravity, "all they doing all day is *surfing!*" No sooner had the towns they spoke for passed than these messages started breaking up, occasional phrases lingering on—a girl's name, an unlikely dollar amount, a reference to Corinthians or Waikiki Beach or some peanut wholesaler—before being chased from the frequencies entirely.

America was selling God, urinary tract-friendly cat food, the clothes of departed fathers and much else besides. Thickets of hoardings preceded each interstate exit, promising fast food, budget beds and cheap gas, factory outlet stores, furniture warehouses, state forests and parks, and museums. Something billing itself as Georgia's "Famous Southern Burlesque" caught my eye. That had a mannerly ring. What is actually meant became increasingly clear as the attraction approached. "Daringly bare," read the sign, less coy and rather more commercial, at the penultimate exit. And, approaching the actual exit, as if to dynamite any lingering misconceptions: "We're butt naked!" above a large, pink silhouette of an impressively benippled cowgirl in nothing but a Stetson.

Far beyond the nipples, I left the interstate and headed into Georgia's south-west. Mashed corn stalks and scraps of white cotton littered fall-harvested fields, abandoned battlegrounds now. Ponds lay treacle brown below red Dutch barns. Crows gathered on phone lines to cast long shadows. After the worldly temptations of the interstate, offers of redemption abounded on the backroads. "Try our salvation; it's free," declared the sign outside a Baptist church. "Jesus paid the price." Many of the signs outside old weatherboard homes and country stores—"erti rakes" or the soliloqual "to be"—had shed in clumps too many of their black-plastic characters to make immediate sense. In time, however, I would come to be proud of my decoding skills, recognizing these retrospectively as *certified brakes* and *pinto beans*.

In the afternoon, I stopped for fuel at a single pump. Nobody was about. An interested horse watched me from across the road. A breeze, an event in itself, came from nowhere to set an unseen outdoor mobile tinkling. The horse pricked its ears. The solitary girl inside the store leaned forward on her elbows to check the sky for signs of weather.

"For the petrol," I said, pushing a ten-dollar bill between the towering stacks of Lucky Strike cartons that framed her. "Sorry," I smiled. "The gas."

"Love your accent," said the girl, picking at coins in the till like a monkey flea-grooming her mate. In the backroom, a clock ticked heavily, dragging itself towards the girl's shift end. Her hair was cutely bobbed. A yellow butterfly-shaped hairclip had alighted on her forehead, keeping her hair from her undeniably sweet eyes; dammit, I thought a starlet from Scott Fitzgerald had crash-landed in a setting from William Faulkner.

"Thanks," I said, casually. An Englishman in the South, I had been advised, could expect to hear such things.

"It makes me wet, your accent does," said the girl. *Way-ut.* She dropped the coins, one by one by one, into my outstretched palm. Plain talking. I could now see why *Famous Southern Burlesque* had lost out to *We're Butt Naked.* "It makes me married," was how I tried to reply, thinking that to parry her on the grounds of circumstance rather than preference was only polite. What came out, as I backed from the store, were incoherent utterances and placatory gestures, hands half-raised. A passing driver, an unlikely event on this lonely road, might have imagined I was about to take a shotgun blast to the torso. It was only when I got to the car and looked back to see her staring into the distance that I realized it had been nothing personal. You say that to all the boys, I said to myself. I bet.

"I'm heading for Climax!" shouted a smiling John Calhoun the following morning. He was a short man in big boots and a battered, black pick up, with a country band implanted among his vocal chords. Utterances sprang from him, full of pluck, twang and soaring background notes so that they sounded less like sentences than song lyrics. And now he was asking me to ride out with him, out to a place called Climax.

We had just met in Whigham, a town of a few hundred souls housed in peeling clapboard homes loosely arranged around a single traffic light. Whigham was a few turns off the highway, some grain silos, a hardware store and a United Methodist church, a gas station and the general store where they did hot dogs and coffee. Whigham was an hour's drive west of the girl who loved English accents, a few miles north of the Florida border and just east of Climax, a small town where you could stand on the main street and just watch for the sniggering from the passing cars.

"It's a wonder they don't ee-rect a sign at the city limits reading, 'You Are Now Approaching Climax,'" said Calhoun, installing the sign in his imagination with a flourish of his right hand. "Might make a purty fine joke! Stead there's all those folks think them Climax citizens just too durned upright for their own good. Like all they got over there is that Swarn Tarm."

Swine Time, it transpired, was the Climax Community Club's prize pig day on the Saturday after Thanksgiving, which Calhoun clearly didn't much rate.

"Mind you," he added admiringly, "they sure got some mighty fine rattlers over their way." Calhoun had a friend owned a service station. Also owned the land out towards Climax where we were going.

"Two hundred and fifty acres," he said. "Hunted it every year for the last eight. Ain't never been disappointed so far."

Calhoun was a carpenter. Built wooden churches, which made for a pretty good living.

"In these parts," a contented Calhoun told me, "there's always churches need building." Today, however, the latest church would have to wait. Calhoun was taking a day off to hunt eastern diamondbacks, America's largest species of rattlesnake, in what was considered some of the best remaining snake country in the south-eastern United States.

"I'm telling you straight," said Calhoun as we drove among crop fields fringed by plantations of pine and stands of wireglass. "Ain't no rattler country comes much better than this." The crops, he explained, attracted the rats on which the snakes fed. And the adjacent sandy pinelands were riddled with the burrows of gopher tortoises in which the rattlesnakes liked to lie up.

"Like havin' your bedroom next to the kitchen," Calhoun explained. He turned off the road, nosed the pick-up between gate posts and came to a halt on a bed of pine-needles. There was a frost; that was good news for me. But where the sun fell, steam was already rising. That was more worrying.

"Nice boots," said Calhoun as I climbed from his pick-up. They were indeed nice. And made of the thickest leather I had been able to find.

It was the day before Whigham's annual rattlesnake round-up, one of a hundred-odd such events to take place all over the United States, but mostly in Georgia and Alabama in the east and Oklahoma, Texas and New Mexico out west. The rattlesnake round-up is an established part of the American rural scene. Americans were flushing out rattlers in Iowa as long ago as 1849. It had become a community event, an annual observance of the country calendar that invited broad and various comparison with a whole range of similar rituals worldwide: the slaughter of pigs in Spain (the matanza) or of sheep in the Islamic world, for the round-up was in one sense a harvesting activity (of snakeskins, of meat and even of gall bladders for Far Eastern males looking to pep up their virility); the Spanish bull-fight or the Masai lion hunt, for it was also a demonstration of skill and courage, or so the participants would have you believe; and, finally, it compared in its fundamental aspect at least with the British fox hunts, for the round-ups were also considered a service to the community, ridding the land of unwanted creatures.

Like most rattlesnake round-ups, the Whigham event began as a cull. Back in 1960, a number of local quail-hunting dogs, plus the occasional human, had got themselves bitten by rattlers. Rattler numbers were thought to be getting excessive. The round-ups, which soon started attracting the curious, rapidly evolved into a lucrative local event. The area's motels offered rattlesnake weekend packages. The town's gas station put a caged rattlesnake out front and posted a bold, felt-tipped DANGER sign beneath it. The round-up programme attracted many congratulatory ads from local business (which, notably River

Bend Ford Mercury's "Where Ya'll Always **Strike** Your Best Deal," were mostly predisposed to the direst snake puns). Hundreds of street vendors came in for the round-up. And the participants in the town parade—the fire department, the sheriff's department, local senators, church groups and the Whigham Jaycees—had been practicing for weeks. And for Little Miss Whigham, who was dressed in sash and Stetson and rode past on a trailer piled high with hay-bales crawling with inflatable plastic rattlers, it was about the biggest day of her life.

There had been a time in the northern states, where the snakes tend to "den" together for warmth, when round-ups produced spectacular results. Back in 1849, two Iowa men caught ninety rattlesnakes in the same number of minutes. But in the mild south-east, where snakes do not "den," John Calhoun could spend hours securing a single rattlesnake. For weeks now he'd been catching snakes for the round-up, and had maybe twenty in boxes at home.

"Don't expect snakes everywhere," Calhoun told me. "But they might just be out sunning themselves this morning and I'd advise you not to tread on one. Not even in those handsome boots." Calhoun clearly found my boots amusing, but it wasn't that which upset me. Live, wild snakes were no longer a distant prospect, and I asked Calhoun whether he'd ever been bitten.

"Funny how jus' 'bout everybody asks that question," he replied. "But no, the answer is I ain't."

Calhoun disappointed me. People who spent their lives going after hazardous things were supposed to have run-ins with their quarry. Otherwise, they were either just too damned good to be true, or the prey wasn't so hazardous after all. Shark hunters were supposed to be able to show you the shredded wetsuit that had saved their lives by holding their chewed torso together. Men who went after lions should be able to describe what it felt like when lion molars closed upon the skull, then went loose as the life finally left the bullet-riddled beast that had attacked them. And you were tempted to ask what a bear hunter had been doing if he had never been treed by *ursus horribilis*. Instead of which Calhoun was hauling a fifteen-foot section of ¾-inch black plastic piping off the back of his pick-up, and lecturing me on gadgets.

"See," he explained. "I've plugged the end of the piping with a twelve-gauge shotgun shell, and punched some holes in it a little ways up. You'll see why." He made a mental check of his gear: a clip-top bucket, a hand mirror for reflecting sunlight down holes so he could see within, a small bottle of gasoline and a snake hook shaped like a shepherd's crook.

"Made from an old golf club," said Calhoun proudly. "With a ⅜s iron rod welded to it." He made to leave, patting the empty rattler boxes on the pick-up as if to promise them contents by the day's end, and set the padlocks swinging with his trailing fingers.

I checked out Calhoun's own boots, and asked him whether he ever smeared them with the musk of in-season female rattlers. Some hunters used to do that, I had read; that way you could quickly attract a lot of horny male rattlers.

"They did what?" said Calhoun, sounding disgusted. He wandered off into the woods, trailing the length of piping behind him like a lost utilities man. When he reached a hole, he shoved the piping down it, stirred it around and then put the available end to his ear.

"Listening for rattlers," whispered Calhoun menacingly, and Ry Cooderish guitar notes sounded across the still pinelands. Or should have.

"If he's down there, I'll hear him rattle. Doesn't like the intrusion. You know," he continued conversationally, "I'm thinkin' of getting me a vibrator. Rattlers hate vibration, see. Stick a vibrator down a gopher hole, I reckon, and stand back."

"Where would you get one of those from, John?"

"Oh, you know," he replied coolly, scanning the landscape for fresh gopher tortoise holes. "Magazines. Books. That kind of thing."

All morning, we walked among pine trees and pecan orchards under a bruised winter sky. All I saw, of course, were a lot of fallen pine-needles; despite the boots, I tended to keep my eyes firmly on the ground. Calhoun, I couldn't forget, had brought me here because there was no better place for finding snakes. I needed to be sure where I was placing my feet. The real danger, therefore, was of bumping into trees. Calhoun, I noticed with some embarrassment, was allowing himself a whole series of inward smiles at my boots. I think I'd over-dressed.

Calhoun awoke his first rattler in the early afternoon. Suddenly, he was alert, one forefinger lifted in signal as he screwed the piping hard into his ear.

"Got ourselves a rattler," he whispered. "And now we gotta get him out."

It was then that the plastic piping, which had already functioned as stethoscope and prodder, began to demonstrate its true versatility. First, Calhoun attached anglers' hooks to the end of it, so it operated as a fishing rod, then flexed it in the hole. But he could not hook the snake. After several attempts, Calhoun laid down the piping.

"OK," he whispered. "Let's drive him out."

With that, he took the bottle of gasoline from his pocket and poured an amount into the piping so it now acted as a funnel. He put his mouth to the piping, gently blew the toxic fumes into the hole and stood back. Gassing rattlesnakes had recently been made illegal in Georgia, but gassing rattlesnakes happened to be effective. Within seconds, an eastern diamondback rattlesnake was appearing from the hole.

Involuntarily, I stepped back. My body stiffened and I simply forgot to breathe. For some time, nothing but the snake existed. All awareness of the winter-brown woods, of Calhoun and my boots was lost as the snake overran my consciousness. Its head was broad, and slung below the neck like a heavy, sinister pod. Behind the head trailed an ivory-toned diamond design, edged

with black on a dark-brown background, running the length of
the snake's emerging body. It was a heavy creature, its body rising to the suggestion of a keel.

But Klauber had been right; the snake was in no state to
harm me. When Calhoun slipped the hook under its midriff,
the rattle that was said to set a coyote's sphincter quivering at
fifty paces had all the conviction of a clockwork toy expiring.
The snake hung sluggish from either side of the hook. Only its
head and rattle showed faint signs of movement. They were upturned at the extremes, lending the snake the overall shape of a
vaudeville moustache. By the time Calhoun had dumped the
snake into his bucket, a second snake was emerging. This one
was weaker still. With his hook, Calhoun raked it clear of the
hole, so casual he might have been turning steaks on a barbecue grill. And all the snake could manage in the disabling cold
were single clicks of its rattle.

"Can you pee?" asked Calhoun, standing over it.

"I'm sorry?" I replied.

"Pee," he repeated. "Can you pee on the snake to bring it
round?"

"I'm sorry," I told him. "Just been."

"Well, let's see," said Calhoun, turning away from me to
undo his fly. He finally summoned up a stream of warm piss
which he directed at the snake, nailing it on the head. Being
pissed on revived the snake, and was probably a greater source
of annoyance than being ejected from its hole. It put up an impressive rattle and even attempted to strike, but this piece of
vaulting ambition turned into a feeble forward slump that
Calhoun had no trouble avoiding, despite his undone fly.
Calhoun casually slipped the hook beneath the snake's belly. It
hung limp, like an oversized and brilliantly decorated pork
sausage, the sort of exhibit that might have gone down a storm
at Swine Time.

Was this the "hedious serpent, the formidable Rattle Snake"
of the early settlers? The creature that could fly, poison with its

breath and grow to ten feet or more? The creature that, "by the rapidity of its motion, appears like a vapour"; that "continually menaces death and destruction, yet never strikes unless sure of his mark"? How come, then, ours had just struck with all the accuracy of a drunk jawing a lamp-post?

I was beginning to see why John Calhoun had never been bitten.

Darlene stood above the second box, praying hard. Then she gradually moved her good hand and slid it under the larger of the two rattlesnakes in the box. It was a mild October night, and the snake was alert. At her touch, it peeled off a tight drumroll of rattles and withdrew its head sharply. But it did not strike. The texture of its skin against her palm reminded her of warm, dry varnish, and she remembered how good snake-handling could feel. Then she felt Glenn's hold on her hair loosen.

"OK," he said. His voice was hollow, perplexed. "I'm going to let you live since the Lord let you handle that one." Glenn was impressed. All drunk and mussed up, Darlene had just handled the meanest of all his snakes. It was like the best of times in church, when God turned the strike aside and you could just feel the victory all over you. He hadn't expected it that way. For now, all the murder in him was gone. He felt uncertain how to proceed. In the end, he led her outside and walked her to the Chevy, even lending an arm for support. They sat in the car, listening to the night-birds in the woods, and for a long time neither spoke.

Darlene was beginning to feel real bad. The pain around the bite was extreme, hot as a hob you'd whip your hand off if only you could. The whole left hand looked swollen now, the skin stretched in an unfamiliar way. She felt nauseous and dizzy. Shapes were appearing in front of her eyes, little glittery yellow lozenges that fluttered like falling leaves. She was also very thirsty, but she guessed that was down to the alcohol. Her tears broke the strange, becalming spell in which they had found themselves.

"Glenn," she whispered. "You don't want me to die down here?" The question had trace elements of assertion. Glenn helped her from the car and led her back through the woods. But she stumbled on the way, and felt so weak she couldn't get to her feet. Glenn went ahead to fetch her a glass of water from the bathroom. In the darkness, she became fearful that he had left her there and she dragged herself to the house. She passed out on the porch. The next thing she knew was her husband standing above her, kicking her in the ribs. The calm was over.

"God told me you was going to live," he screamed. And he kicked her again.

"That truly scared me," Darlene would recall. "I knew he didn't have nothing to do with God that night, you know, being drunk and mean and saying things about God."

Finally, Glenn got her into the house, laid her on the couch and brought her a glass of water. As she lay there, great pulses of pain clamping around her thumb, she became aware of wet patches spreading against her skin. Then she noticed the smell, and realized that her husband had pissed over her as she lay unconscious on the porch.

∽

Whigham's big day broke bright and warm, with cotton clouds strewn across the sky. The promise of good weather had brought out the crowds. There were Alabama, Florida and Tennessee plates among the station wagons and sedans that were stacking up along the roadside. The approach road to Whigham's school was lined with stalls selling burgers, ice-creams, Polish sausage, bird houses and pet accessories, arts and crafts, Stetsons and flags, and snake products. There were snakeskin belts and wallets, snakeskin scabbards and boots. There were earrings made from rattler fangs and necklaces from their rattles. There were snakes on T-shirts and there were inflatable plastic rattlers, edible candy snakes and even rattlesnake fritters.

Either side of the high wire-mesh fence that enclosed the pitching area of the school's baseball diamond, the organizers had positioned cattle-gates to create a corral for the round-up. Large wood-framed pits, with transparent Plexiglas sides, had been set up. The first arrivals were already gathering beyond the cattle-gates, staring at the pits where a few rattlesnakes lay, mostly diamondbacks and a few chevron-patterned canebrakes.

With a due sense of theatre, most of the snake hunters chose to delay their arrival until the crowds had thickened. This was their day in the sun, and they were set on enjoying themselves. They'd done their hair this morning, had these guys. They were eventually chauffeured into the corral on the backs of their pick-ups, where they perched amongst the old ammunition boxes, fertilizer containers, crates and even the occasional burlap sack that contained their haul. As they clambered from their pick-ups, they acknowledged the milling crowd with lazy waves. They were like bounty hunters from another age, turning up at the local jail with a bunch of desperadoes for the cells. These men dealt with rural America's Public Enemy Number One. Put their lives on the line to make the land safe. So folks could sleep safe at night. It was God's good work. Didn't the Bible, Genesis 3, say of the snake that the "seed of woman shall bruise its head"? Well, these guys didn't stop at head bruising. They'd cleared the area of six hundred plus snakes last year.

"You ever been bitten, mister?" a young boy asked a tall snake hunter from the other side of the mesh wire. Through thick-lidded eyes, the boy's mother looked on, thoroughly impressed.

"Nope," the hunter replied. But his tone suggested an unspoken "yet," as if even he wondered how he had got away with it for so long. "Let's hope my luck holds, buddy."

But I knew otherwise. I had seen John Calhoun piss on rattlesnakes. The truth was that winter temperatures so improved
the odds in the snake hunters' favour that this supposedly intrepid activity was no more dangerous than pest control. Sure,
some contests between man and beast exposed him to a degree
of risk—stalking African buffalo, nineteenth-century whaling,
even rodeo-riding—but rounding up deep-frozen rattlesnakes
could not be counted among them. A closer comparison might
be seal clubbing. Except, in the round-ups, there was no danger
of going through the ice. To suggest, however, that these men
were less God's foot-soldiers risking their lives to serve the community than the chaps from Orkin might not have endeared me
to the people of Whigham. So I looked suitably impressed and
sneaked under the cattle-gate to get a closer look.

The snake hunters worked in teams. As they delivered their
hauls, numbers were counted, and exceptional specimens
weighed as possible contenders for the cash prize awarded to
the year's heaviest snake. A dealer from Florida, who would
turn the snakes into boots, wallets and pet food, was paying $6
a foot. As his assistant held up each snake by the tail, the dealer
called them in feet. "Three and a half," he offered. Invariably,
that was the signal for the hunter responsible for the snake to
scowl, mumble how he had trudged miles for that miserable
critter—judging by his fancy numbers, the dealer was not wearing holes in *his* boots—and finally counter with his own estima-
tion of the snake's length: four feet. "Yeah," exclaimed the
dealer derisively. "If you stretched it till it snapped."

On the other side of the enclosure, a man called Ken was
also handling the snakes. Ken was overweight, and wore a
beard and red braces. By the time the snakes reached Ken, the
heat in the Plexiglas pits had made them active and irascible,
their rattles fizzing like cicadas. Ken manufactured antivenoms,
and was "milking" the round-up's every snake. With his hook,
he pressed upon each snake's head, grabbed it around the neck

and forced its fangs to bite into a rubber flange stretched over the rim of a funnel so the yolk-yellow venom ran into a collecting vessel packed round with ice. It was a process Ken would repeat hundreds of times during the day. Strong hands were vital. Still, Ken found time to play to the crowd between milkings. Clamping the snake's lower body between his flank and elbow, bagpipe-like, and holding the snake by the neck much as one might support a hand puppet, he worked his way down the fence-line. In the barrage of utterances, exclamations and expressions of profound distaste that his passing prompted, the same question was repeatedly asked.

"You ever been bit?"

No, said Ken. He hadn't. Wasn't intending to neither.

Ken made antivenoms. Ken had a pretty good idea what a bite from an eastern diamondback, the most dangerous of the North American rattlesnakes, could do to you. Ken had seen the blisters black as olives, big as plums. The forearms cut open to the bone to relieve the pressure. Ken was a scientist. Ken wasn't planning to get bitten.

But, boy, did people ask. That day, the same question that I had asked John Calhoun rang out around Whigham a thousand times. "Ever been bit?" sounded round the hunters and the snake-pits. It was etched into the sound waves around Ken. By dusk, you had heard it so much you might even have expected it to arc, in a fetching pink, across the sky above the town's little restaurant, The Old South.

There was nothing smart about getting bit. The snake hunters knew that. Still, everybody wanted to know what it felt like, and the fact was the hunters couldn't say. They were the snake men and they didn't know. Sure, not getting bit was the sensible option, but taking a bite and surviving it, why, that was something else.

In its thirty-six years, the Whigham round-ups had recorded only one case of snakebite when a visiting herpetologist demonstrated the speed of a striking rattlesnake rather too spectacularly for his own good. But out west among the Oklahoma round-ups, out at the little town of Waynoka, they even had a club for snakebite survivors. They called it the White Fang Society. Unlike my club, a contrivance that existed only in my mind to provide me with a psychological lift, a means of getting me through the nasty business of meeting snakes, the White Fang Society actually existed.

Not that it could be called an active society. Membership hardly conferred any material benefits. The society never even met; it had no business to speak of, and the few members knew each other just fine as things were without making a special occasion of it. Membership was no more than a certificate posted on the wall of the hunt club on Waynoka's Missouri Street. But everybody knew what membership meant. A bite from a rattlesnake was less an injury than an initiation. It was a rite of passage that put you at one with the land. Marked you out as a frontiersman. Surviving a bite showed you had been there. Taken all that this great land could throw at you, and endured it with no more fuss than a grimace. There were plenty out there, I reckoned, who would give anything for a rattlesnake bite. You could wear a rattlesnake bite with pride. It was rural America's ultimate status symbol.

Frank Dobie, who extensively chronicled rural American life earlier this century, understood what a rattlesnake bite signified when he wrote the story, most likely apocryphal, of a Texan girl called Dolly Dickens. It so happened that each and every member of the Dickens family, except for Dolly, had suffered a rattlesnake bite over the years. Dolly might have been grateful to have escaped such an injury, but she increasingly came to understand that she should regret not having been bitten; it somehow made her incomplete. The other members of the family, wrote Dobie, "had the clan spirit and were exceedingly proud of their record." That her skin was unbroken by the fangs of a rattlesnake became a source of increasing humiliation to Dolly Dickens. She took to walking about barefoot in long grass. She made a point of not looking where she was going. She courted carelessness. Stuck hands down hopeful holes, which was how one of her brothers had got bitten. One evening, as she followed the cows through a patch of broomwood, Dolly Dickens was finally bitten by a rattlesnake. She ran home, elated. Started doing better at school. Married a well-to-do rancher. In the bite, Dolly Dickens found all the acceptance she had ever sought.

Well, that was Texas in another time, where everybody got bit. In Georgia, things weren't quite the same. The year's heaviest snake, at 9 lb. 5 oz., had broken Terry Harrison's snake hook, but, no sir, it hadn't bitten him. Hadn't never been bitten, had Terry. An elderly man who appeared continually to be chewing on air must have overheard me.

"Fellow up Enigma way got a bad one last year," he said. "One of those crazy 'ligious fellas. Killed him. Yeah, those guys are always getting it."

I assumed I had not heard the old man right.

"Those 'ligious folks," he repeated. "Holy Roller types. Handle snakes in church. Get bit. Die sometimes. As I said, like the fella out at Enigma."

My first reaction to the first of the many snake-handling deaths I would hear of; *the fella sure lived in the right place.* This was modern America, and I was now hearing of people who handled snakes as a rite of Christian belief, regularly took bites and refused medical attention to boot. I had heard something of the practice before but, assuming it had long since died out, had mentally consigned it to the closed doors of history.

"Doubt they'll talk to you though," said the old man. "Keep to themselves, those folks."

I would see. Enigma was two hours to the east, a tiny settlement in Central Southern Georgia. I set off the next morning, detouring the girl at the gas station just in case. A glittery frost was fast turning to thaw in the bright sunshine and my mood was high. It was Sunday, and the parking lots outside the Baptist churches gleamed with rows of clean, shiny automobiles. On the radio, a preacher was grappling with an extended food metaphor.

"Our children won't eat plain cereal," he hollered. "They've got to have some other kind of cereal that's already doctored up. It's got to have sugar on it, or honey on it, or nuts on it, and even then sometime we throw in fruit and stuff of this nature to, sort of, doctor it up." I took him to be talking about the way we complicate our lives, losing sight of simple needs in the process. Still, that did not warrant such an outspoken attack on muesli.

I drew up outside Patricia Hale's home. It was a cream-coloured, weatherboard cottage, one of a number arranged along a cul-de-sac, with a neat garden to the front. On the swinging seat, suspended by chainlinks from the eaves above the verandah, a comfortable cat was curled. There were shrubs and a carport to the side. When the door opened, the gospel music reached out to enfold me but Patricia Hale held back. I told her why I had come. She invited me in, but not without misgivings. "I'm just back from church," she told me. She wore her long hair in a strict bun. She dressed austerely and her face was disfigured, which had the effect of dramatizing her grief. She directed me to a seat and sat opposite me, a Bible positioned on the low table to her right.

A few days before, Patricia had commemorated the first anniversary of the death of her husband, Bruce. Bruce had been bitten by a rattlesnake during a Sunday morning service at the couple's church, the New River Free Holiness Church in peanut country just a few miles to the south.

"It was on the forefinger," said Patricia. "I think maybe he took two bites. Must have been about midday he took the bites," she remembered. "For a while he just carried on singing and holding them snakes like nothing had happened. Oh, glory. But after a while, you could tell he was feeling it. Sweating,

wincing and slurring and stuff. He didn't want to go home. He
lived for church did Bruce. But he was turning sick. So we put
him in the car and drove him back here, myself and our
preacher. And other friends and family, they come along too.
We put him on the couch and spent the afternoon in prayer. We
prayed hard for Bruce, we did, and for a while he seemed to be
holding up OK. There was swelling up the arm, sure there was,
but nothing too serious. Not till the light started to go when he
got suddenly weaker. The swelling got worse and he began to
sweat heavily. He was suffering hard."

"You never thought of taking him to the hospital?" The most
obvious question was also the cruellest one.

"Bruce didn't want it. He put himself in the hands of the
Lord." Patricia shrugged. It was as she would have done.

"We don't handle snakes often as we used to in these parts,"
said Patricia, "not like they do up north in the Appalachians.
North Georgia and Alabama. Tennessee and Kentucky. But
sometimes, when the spirit moves on us, the men still bring out
the boxes. Matter of fact, they was handling only today. And,
me, if I'm moved to do so, I'll handle them snakes. If the Lord
wants me to."

Surviving snakebite meant something different here; what
was a kind of initiation, a blooding to most rural Americans,
was no less than divine acceptance, evidence of the Lord's protective hand, to the Holiness handlers.

But Bruce Hale had not survived the bites. He died at nine-
thirty that evening, suffering to the end.

"Some people say my husband's faith was not firm enough,"
Patricia explained. "That's why he died, they say. But that
doesn't mean he wasn't the best kind of Christian man."

I wondered what effect his death had had upon her own
faith. She reached for her Bible, wrapping fingers round its
dark spine.

"I hope and pray my faith's right good," she said. "The Bible
speaks of famine and pestilence prior to the end, and that's

what we got now. In India and Africa. The end ain't far. I just know it's not long before the Lord calls me home."

Patricia offered to show me her husband's grave; from there, I could find my way back to the interstate. I followed her car down backroads to the New River Church, a solitary modern building set back from the road and enclosed by fields and occasional stands of pine trees. Patricia had arranged fresh flowers—yellow roses and red carnations—at Bruce's grave only that morning. The grave was arranged like a double bed: Bruce had taken the right-hand side. The covers were a grey scattering of marble chips. On the headstones, which doubled as headboards, Bruce and Patricia's names and dates had been inscribed upon the pages of an opened book. Between the headstones a cross stood, with the words "Glory to God" inscribed beneath it. There was no reference to the manner of Bruce's death.

Patricia pointed out the resting place of Leon Parson, another handler who had died from snakebite received here in 1972, aged thirty-two. Parson's photo, in which he wore his balding hair cropped and was dressed in a dark, checked shirt, was inset in an oval plaque on his headstone. His wife, Marie, had still not joined him.

I had heard it said that the probability of an American dying from snakebite was about one in three million. Yet here I was, looking at two snakebite deaths in a single graveyard. I had come to America for heroic snake hunters who, it transpired, never got bitten. And found something else, shy snake-handlers who were forever taking bites and choosing to die in the Lord.

As I drove away, Patricia was still standing over her own grave. She looked ready for bed.

9. AUSTRALIA III

There is generally no lack of courage in the inhabitants of the
Australian bush, but it runs in the wrong channel, and often shows
itself in chopping off the wounded toe or finger—a very foolish and very
dangerous thing to do.

—Gerard Krefft, *The Snakes of Australia* (1869)

OCTOBER 1996

At two o'clock, as advertised, a man in khaki shorts and boots
clambered into the snake-pit. He was carrying two canvas bags,
a bandage and a metal snake hook.

"Folks," he announced. "My name's David and I'm here to
tell you something about some of the one hundred and fifty
kinds of snake here in Strayer." The snake-pit had sheer concrete sides; David seemed to be addressing his audience on the
concrete bleachers from the floor of an empty swimming pool
in which an occasional sapling had sprouted.

Still shaky after my run-in with the seagull, I had left Cairns
that morning, and driven north through a green, field-strewn
littoral that gave way to a glittering ocean. Eucalypts overhung
the road which wound above a deserted rocky beach. Twenty-
five miles north of Cairns, Hartley's Creek Crocodile Farm appeared by the roadside.

"Folks," David continued. "It may be the case that here in
Strayer some two-thirds of all our snakes are venomous in some
way, shape or form, and we have twenty of the world's twenty-
three most venomous species, and all of the ten most ven-

omous." The crowd stirred in anticipation at the impressive statistics.

"But, folks, don't worry; no snakes are in the business of slithering up and biting just about anybody they can. Leave them alone, and I can pretty much guarantee they'll leave you alone. In fact, on average, just one point eight folks, folks, die from snakebite every year in Strayer; a negligible danger. Folks, let's talk about snakes in a sensible manner for a change."

I could feel the folks either side of me going down like aged party balloons. These people had not come to the daily snake show at Hartley's Creek to be told that Aussie snakes constituted a "negligible danger." They did not want safety guarantees. They had not expected the "sensible" approach towards what David might have termed the local herpeto-fauna. In their view, David was in danger of castrating the great Australian outdoors. And Australians, who have tended to take a perverse pride in the dangers lurking amongst their wildlife, weren't going to stand for that. Call this country safe, they would be sure to retort, when our coastal waters have some great shark attacks to their credit. Several of our spider species have potentially fatal, not to say agonizing, bites. And there have been some impressive croc attacks, some no distance, they would doubtless point out, from Hartley's Creek. So less of the sensible nonsense please, David, and let's get on with the snake show.

They had a point. The Australian snake show had been around for centuries. It had always had a melodramatic character, and to peddle anything other than blood-curdling yarns of death-dealing monsters was, frankly, to misrepresent it willfully. The snake show was about thrilling yarns; the unfinished, scrawled note that read "Bitten by a..." lying beside the body, the snake man's derring-do experiences in the Pit of Death, the unprovoked attack, the slashing fangs dispensing enough venom in one strike to kill a hundred men. Such had been its time-honoured format ever since the 1820s, when a man called Wilkinson wandered around Parramatta, Sydney, with a bag of snakes. When the mood took him, Wilkinson would stuff the bag under his hat.

Wilkinson had no use for snake hooks. He is said to have caught his snakes, often deadly ones, by approaching them from behind, seizing them barehanded and putting them into his bag. His party piece, often requested of him, was to plunge his hand into the bag, and he never seemed to suffer any subsequent ill effects.

"A lot of people out there generally have fears and phobias about reptiles, folks," David persisted. "But use your intelligence; these creatures are pretty much more scared of you than you are of them." Into the Pit of Death, as any self-respecting snake man had once termed his show premises, the dead hand of Common Sense had crept. It was apparently the snakes, not the humans, who had come to be fearful.

"Now, folks, that's not to suggest you shouldn't be prepared. So just in case a very unfortunate incident should happen somewhere, I'll be giving a first-aid demonstration so you'll know exactly what to do."

Which is more than could be said for Australia's early settlers. Popular nineteenth-century cures for snakebite included gunpowder, which was heaped upon the wound and then ignited. Bite victims were known to remove the bitten extremity with an axe or even a loaded gun. Leeches, on account of their sucking ability, were often applied to bites. On the principle that the application of a second poison would neutralize the first, strychnine, mercury, ammonia and "chlorine of gold" all enjoyed spells of popularity. The juice of banana leaves had its supporters, as did the urine of toads or even immersion in cattle manure. And massive quantities of alcohol, even more than was generally downed as a daily salve against the brutish realities of nineteenth-century Australian life, were regarded as an excellent cure for snakebite. A quart of brandy by the glassful, one every ten minutes, was a widely accepted prescription. *The Australian Medical Journal* of 1859 quoted a memorable case of snakebite treatment in which the wound was "blasted with gunpowder." The victim "was quite ill for eight hours, during which she was treated internally and externally with ammonia, cold effusion constantly applied, and as well drank two bottles of brandy. She then rode home on a man's saddle."

Not all Australians, however, were quite so robust, and by the mid-nineteenth century many were clamouring for a rather less elaborate treatment for snakebite. A range of specific snakebite nostrums and potions were concocted accordingly and the snake shows of the period were effectively little more than poorly disguised pharmaceutical product demonstrations. But the shows were highly dramatic. There were considerable profits at stake, and the competing antidote salesmen were prepared to go to extremes to convince the public of their potion's relative effectiveness, even if it meant putting themselves at extraordinary risk. Wilkinson of Parramatta's behaviour may seem reckless, but subsequent generations would stage far wilder shows than his before David and his kind were to reinvent the snake show as a thrill-free lecture one hundred and seventy-five years later.

In Tasmania, where the Governor's wife had offered a shilling's bounty on any snake brought in during the 1830s, an early snake man called Joseph Shires regularly accepted a bite from a snake before a paying audience and then applied his antidote—a closely guarded secret formula—as proof of his potion's efficacy. Once, Shires took bites on the finger, lip and neck from more than one Australian tiger snake—considered the world's fourth most dangerous—before collapsing and being taken to nearby lodgings, but only so as to recover in time for his next demonstration. In the early 1900s, fliers were regularly distributed around towns inviting residents to forthcoming "demonstrations." "Young Milo," read one, "will allow a venomous snake to bite him and will then treat himself. The snake will immediately afterwards kill a rabbit." Now *that* is what most Australians would call a snake show.

Repeatedly, potions such as Morrissey's Snakebite Antidote seemed to publicly substantiate its claim to have "cured without failure every case of snakebite to which it has been applied in Australia." No surprise, then, that it sold successfully throughout the country. It and leading rivals such as Underwood's Antidote for the Bites of Snakes and Other Venomous Reptiles became legendary brands, and attracted extensive speculation as to their contents. In his *Bush Wanderings of a Naturalist* (1861), Horace Wheelwright wrote that Underwood "discovered the secret of his elixir by watching a battle between a snake and a stump lizard. After the lizard had killed the snake, he saw it eat the leaves of a small plant"; presumably, the lizard was treating itself against a bite it had received during the battle. So Underwood "gathered some, made a decoction of it, and this was the secret of Underwood's mixture."

As late as 1916, the renowned Tambo the Reptile King drew over 4,000 paying visitors to one of his Sydney demonstrations. The details of such demonstrations varied, but the basic plot remained the same; chickens, rabbits, dogs or pigs were caused to be bitten by snakes. Ideally, the animals were supposed to die shortly after being bitten while the hero of the piece, the snake man, took multiple bites, applied his antidote and survived. And then opened for business at an exorbitant ten shillings a bottle.

Many such snake men took a great many bites. Most of them lasted as long as they did by a happy combination of sleight of hand, good luck and—an alien concept at that time—the acquisition of immunity. The notion that surviving a bite might confer resistance on the survivor against any future bites only began to be appreciated scientifically with advances in immunization research at the end of the nineteenth century. These discoveries not only inspired the first recognition of the possibilities of antivenom—the use of antibodies extracted from venom-resistant blood to treat bites by being injected into the victim—but also provided a far more compelling scientific explanation than the administering of potion to account for the fact that snake men so regularly survived snakebites. For the antidote sellers, these discoveries signalled the beginning of the end. As early as 1869, Gerard Krefft, curator of the Australian museum in Sydney, was claiming that "the antidote vendors and their supporters" had been "thoroughly exposed." Of course, scientific scepticism took time transferring to the public mind and some antidote sellers would continue in business well into the twentieth century. But many Australians, who had spent good money on what turned out to be ordinary ammonia, vinegar or some other patently worthless concoction, would not be fooled again.

The antidotes and their peddlers may have disenchanted Australians, but public fascination with snakes proved more durable. Towards the end of the century, the snake shows began reinventing themselves accordingly, with the snakes themselves slithering centre stage. The shows were transformed from spectacular pharmaceutical sales pitches to entertainments in their own right. The medical format—bites galore, dead animals and the administering of the antidote—gave way to thrill and theatricality, a danger-steeped vaudeville. In "Sleeping Beauty," a woman would lie still while her accomplice placed deadly snakes all over her body. American performers with names like "Princess Indita" toured the fairs and the carnivals with their rattlesnakes or "Yankee tail-shakers," and performed the famous snake dance of the Hopi Indians of Arizona, which entailed placing the snakes' heads in their mouths. Where the antidote sellers had formerly favoured monikers which attested to their scientific knowledge, most commonly "The Professor," the new breed preferred performers' stage names, often exotic ones. There was "Cobra Boy" and the "Mystery Man." Female performers favoured not only "Princess" but the "Queen of Snakes" and "Wonder Woman." Some even called themselves "Cleopatra" which, given the manner of their namesake's death seemed a willfully suggestive choice and, in some cases, an all too prescient one: a showgirl called Cleopatra died from the bite of a tiger snake near Sydney in 1920.

During the snake shows' heyday, instances of fatal snakebite were commonplace. Plain recklessness, plain deathwish or a bravado willingness to be bitten, often inspired by the mistaken belief that further snakebites could do them no harm, caused the deaths of as many as ten performers in the worst years. Typical was the death in July 1921 of Tom Wanless. Wanless's reputation had taken him on a performing tour to South Africa, sampling the bites of various African snakes, including puff adders and cobras. One Saturday night, Wanless took a full-blooded green mamba bite. The next morning, coughing up blood, Wanless took a look at himself in the mirror, and remarked, "The green mamba wins," before lapsing into unconsciousness and dying the next day.

To the Hartley's Creek audience, it was clear that our David was no Tom Wanless. "Now, this guy, folks," said David, pulling a snake from a canvas bag with his hook in the text-book fashion, "is a red-bellied black. He's not one of the *real* nasty ones. What your red-bellied will do to you, generally, is stop the coagulation of your blood, all right, so you can vomit blood, you can urinate blood, you can also do muscle and tissue damage." One of Australia's most common venomous snakes, the beautiful red-bellied black lay obediently upon the ledge that surrounded the pit. A muscular undertow occasionally rippled beneath its handsome, sheeny scales which suggested, close up, a shingle of overlapping guitar picks, ebony and red.

In his choice of snake—if in no other way—David actually echoed the old performers. The red-bellied black had always been a staple of the shows, the workhorse of the pits, with tigers, browns, copperheads and death adders also featuring regularly. These were the most notorious of the species that were local to the major population centres, notably Melbourne and Sydney, and were readily available to the show people.

"Now, as to the real bad ones," said David. (The absent "folks" lent gravity to his words, and the crowd was all ears.) "The world's eleven most dangerous snakes are all Australian, and of them the taipan, found throughout this region, is one of the worst. Folks, what makes the taipan and a few of the others so bad is that the venom is neurotoxic, which means it attacks your nervous system. Your vital signs go first, blurred vision, tongue swelling, nausea, dizziness, all these sort of things. If you don't get treatment, it's pretty well quite simple: you may die, either from heart failure or suffocation. But due to the first aid techniques we have in Strayer...."

When Kevin Budden died in 1950, Sydney performer George Cann claimed to have been bitten "four hundred times by every type of Australian snake—except the taipan." The exception he allowed spoke most likely for most snake showfolk. It is an accident of geography, and a happy one, that the 1920 show mavericks rarely encountered taipans. Rarely, but not never. Since the species was not widely recognized at the time, it seems likely some taipans masquerading as browns or king browns found their way into the shows' stocks of snakes as they toured Queensland. Contemporary accounts make telling reading in the light of this retrospective suspicion. A "snake charmer" called Mrs. Bydon was fatally bitten in 1927 by what the *Medical Journal of Australia* described as a brown snake at Rockhampton, Queensland—well within acknowledged taipan habitat. "This case," said the *Journal*, "presents several features of interest, perhaps the most important being the demonstration of the fact that, even with the brown snake which possesses short fangs...a lethal dose of venom was inoculated through material of considerable thickness." Or, as now seems far more persuasive, Mrs. Bydon was in fact bitten by a taipan, which looks like a brown but has much longer fangs and an average venom delivery that is thirty times greater.

By the 1950s, however, the snake performers were in danger of being sidelined by stiff competition on the show and carnival

circuit. The world was changing and Cairns's annual agricultural fair, the Cairns Show, was awash with alternative, circus-
style attractions. Nobody could miss the pygmies. "Show Time
is African Pigmy Time," as the *Cairns Post* ad put it. "Little
jungle people—Hottentots and Pigmies, all busy polishing their
arrows, assegais and blow-pipes in preparation for mock attacks
on each other and the audience in general." There was Marg
Van Camp's Pig-a-Dilly Circus ("See the pigs perform very
clever stunts. Opening gates, ringing bells, firing guns...");
Dennis, the skyscraper giant ("The tall man gives a very interesting lecture on his world travels"); Nalda's Doggie Circus
("See and hear them play their Oriental Jazz Band"); Electros
the Human Dynamo ("She will start and operate a washing machine merely by holding the wires between her fingers"); and
Valentina, the amazing invisible girl from Europe. Finally, the
snakes had to compete with Obelia, the million-dollar mer-
maid. Obelia apparently drank, smoked and ate bananas underwater.

The more agriculturally minded visitors to the showgrounds
might instead be drawn to the Guess the Weight of the Bullock
Competition, or to the poultry shows featuring prize breeds like
Anconas, Silkies, Plymouth Rocks and Wyandottes. There were
exhibits of the best strains of sugar cane: Badila, Pindar, Q50
and Trojan. There was a prize for the best plough horses, and
horse races including The Tropical Timber and Hardware Pty.
Ltd. Handicap Trot, and the Garcia Gowns Most Accomplished
Lady Rider (Open).

The snake show was struggling to compete in the postwar
world. It hailed, it seemed, from another time. It had about it
the irreversible taint of decline. History was gradually hauling it
from the present. The emergence of the taipan, however, which
Kevin Budden's death had brought to overnight notoriety, gave
the snake shows a much-needed fillip and proved a prime attraction, if a morbid one. Among the most renowned of the regular performers at the Cairns Show in the early 1950s was

Paula, the Exotic Snake Dancer. "See the Browns, Blacks, A Real Live Death Adder," read Paula's newspaper puff. "Paula is the only person touring Australia today who will show you a genuine specimen of that most dreaded of all Australia's snakes—The Taipan." Unlike earlier generations, however, Paula would not invite any of her snakes, not least the taipan, to bite her.

If the taipan sensation clearly helped the snake shows survive through the 1950s, then not wearing much seemed to be the other significant factor. The snake shows had always exploited the inherent exoticism of their subject matter, but a new permissiveness abroad now persuaded the snake girls that they could go beyond saucy. Sex would replace death as the snake shows' new subtext. Snakes, went Paula's advertising, "creep and crawl all over her unprotected body." Then there was the "incomparably lovely" Melinda Lee who offered, according to her advertisement, "Youth, Beauty, Glamour and Reptiles." Melinda, who appeared at the 1951 Cairns Show, "introduces all the sultry charm of the South in her exotic Snake Dance" in the words of her advertising. "With writhing, hissing reptiles all over her glamorous head and body, this lovely, youthful Tennessee girl displays courage, beauty and reptiles combined..."

But Melinda was no more. (If she had set out on her travels today, I strongly suspected she would have gotten no further than the Famous Southern Burlesque on Georgia's Interstate 75.) The Pit of Death was history. You might hear old Aussie snake men occasionally dusting down their yarns in downtown shopping malls, but the snake show had changed and there was no question of sensible David getting *his* kit off. Instead, he advised us about sensible clothing for the bush.

"Folks," he said, "what you're going to wear is a good set of boots, a good set of thick socks and long trousers that don't cling to your legs in any way, shape or form." He then set about expertly bandaging and splinting a young boy's imaginary bite before bringing out a python which he urged people to touch. He told the audience about snakes' tongues ("sniffing the air, folks, just sniffing the air"), about the way their inner ear structure meant they were deaf to airborne sounds, and signed off by thanking us for our time and recommending the imminent crocodile show.

Among his dispersing audience, however, I noticed a young couple in T-shirts. The girl waited until she was out of David's view to strike suddenly at her boyfriend's chest with a snake's head formed from her right hand, a bulb of fingertips that nipped at his ribcage. All at once he jumped and hollered, stooped to retrieve his fallen sunglasses, and lumbered off after his giggling girlfriend who had scampered off in the direction of the wallaby enclosure. For a moment, her gesture revived the frontier colour, the quintessentially Australian sense of performance—ragtag, melodramatic, irreverent but alive—which had been lost with the demise of the old snake shows. It also demonstrated that snakes remained scary, despite David's best efforts.

I decided to pass on the crocodiles, and drove north on the Captain Cook Highway towards Port Douglas. Cook had had a hard time of this coast, grounding the *Endeavour* on reefs and almost losing her in the process. The name he gave to the headland to the north, Cape Tribulation, aptly reflected his troubles. Two hundred odd years later, of course, progress was easier. The approach road to Port Douglas passed between Day-Glo-green golflinks, hotels fronted by ranks of white flagpoles, country clubs and restaurants with names like Il Pescatore. By the mangrove-fringed harbour, I found the Marina Mirage Shopping Mall, where tourists decanted from air-conditioned buses to exercise their credit cards at Reef 'n Hinterland, Rainforest Images, Kangarucci, a sunglasses shop called Eyedentity, and a pompous trinket outlet called Presenting Australia. A month after my visit, even the U.S. president would turn up here for a few days' golf and snorkelling with his family.

Port Douglas was settled in the 1870s to service the gold strikes inland. It was where prospectors whored, fought and drank, to future prosperity, perhaps, on the way in and whored, fought and drowned their sorrows on the way out, pretty much the way the few who had struck lucky behaved. Once, it had been about the butchest place on earth. So much was evident if you only dropped in on the town cemetery, close by the Mirage Country Club on the edge of town. After the seagull incident, the cemetery had seemed the only place to test my frayed nerves. I mean, if I was ever to go in search of deadly snakes, I at least needed to be able to cope with protective mother birds.

There were no seagulls in the cemetery, but the headstones were instructive. The words "accidentally killed" kept catching my eye. Then, the headstones started getting more specific. There was John More Cole, who died on December 8, 1909 after an accident on the Mowbray tramline, aged twenty-three years and seven months, and Timothy Joseph O'Brien, who was killed during a cyclone on December 2, 1936. I wandered on a bit to discover Sydney Algernon Barnard, who was "killed by blacks at Mossman River on 13th March 1885, aged 23." Then there was William Thomson, who "met his death on 22nd October 1888, by cruel and treacherous murder."

I had a particular interest in the headstones. My other reason for visiting the cemetery at Port Douglas was to find the grave of a local man whose death had been described in the *Medical Journal of Australia* in 1944 under the heading "More Fatal Cases of Bites of the Taipan." The report put the man at two hundred ten pounds and six foot six inches tall, and described him only as "J. P.," but this and the fact I knew the date of death meant I should have no trouble finding him if, as I assumed, he had been buried locally. J. P., who was bitten through the trousers on his right leg at five o'clock in the afternoon, died at ten o'clock the same night. I found him—John Pringle—after a lengthy search. "In loving memory of John Pringle. Died 4th October 1929, aged 44," read a laconic inscription. That Pringle's headstone made no mention of the manner of his death, as if snakebite was a thoroughly unexceptional way to die in these parts, only reinforced the abiding sense that Port Douglas's pioneers did not often pass away in their beds.

That was then, though. Now there were shops called Tropical Desires and Life's Little Pleasures, and a billboard in a patch of bush off the main street which promised that "An exciting tropical shopping village and a 4-star resort hotel designed by Gary Hunt, architect, will soon be developed on this site. 30 retail shops, restaurants and food court, twin cinema complex, 148-room strata fitted resort hotel. Undercover parking."

FNQ was being reinvented. Like the snake show, it had been sanitized, smartened up, rendered sensible until it became unrecognizable, divorced entirely from the spirit of the original. This former frontier settlement, as best an encapsulation of the old FNQ spirit as you could find, seemed to be making a last stand against boutiquification, branded clothing and strata-fitted resort hotels at a) the meat-pie shop and b) the pub in the middle of town, where the sign above the bar read "No Swearing, No Bare Chests, No Dogs." Then I found c) the Snakes Decoder.

It was lurking among the women's magazines, of all places, in the town newsstands, a reminder that FNQ was not just about pesto recipes and tropical shopping villages. It was a fold-out cardboard guide that described the appearance, habitat and, most interestingly, the bite of various Australian snakes (the categories were either Danger! Bite can make you very sick, or Danger! Very venomous. Potentially Deadly). It also included first aid information. The information was laid out in segments, one for each snake, that were arranged to form a circle. There were evidently enough dangerous snakes, eighteen in all, to fill three circles, one on each fold-out panel. The legend and general information appeared on an attached, revolving circular piece of plastic. It had a transparent, segment-sized window which you could turn to expose your chosen snake. It was the best eight dollars I have ever spent.

Things improved further when I tried to walk to the town's famed Four Mile Beach from my accommodation, a series of cabins arranged around a swimming pool.

"No, no," said the breezy landlady as I wandered past in my swimming shorts. "Not that way. That way you'll walk into the crocodiles." I was reminded, as I about-faced, of a local croc attack ten years earlier. A few days before Christmas at the nearby village of Daintree, a group of people had been standing in a mere foot of water, cooling off during a late-night Christmas party. Suddenly, there was what one eyewitness described as "an enormous disturbance in the water—a vortex," and the unfortunate Beryl was gone without so much as a scream, leaving only a ripple behind her.

The beach did not seem much safer. A big wooden board warned against the presence in the water of highly dangerous box jellyfish. On the sign, a flowchart took first aiders through the procedure—check for pulse, liberally apply vinegar (stored by the sign), send for antivenom and commence cardiopulmonary resuscitation. I noticed that the poster was sponsored, with a typically irreverent Australian flourish, by "XXXX: Our Beer." All the swimming went on within protective nets the size of swimming pools. Feeling not so brave, I walked along the shore, dipping my toes occasionally, but only within the netted section. This meant I must frequently retrace my steps since there was limited shoreline within the nets.

That evening, I examined my decoder and listened to a radio feature alerting me to the danger, now that it was the nesting season, from dive-bombing magpies. The Department of the Environment had apparently manufactured magpie-protection hats ($6.50 each) which featured the face of Dame Edna on the back. The idea was that the magpies, like the Indian tigers of the Sundarbans, would not attack a human that seemed to be facing them. Not Dame Edna, anyway. After my run-in with Cairns's killer seagulls the previous afternoon, it perked me up to discover that the local birds were enough of a threat to merit airplay. Perhaps my behaviour in the cemetery had been less craven than prudent. The radio feature also made me feel better about Port Douglas. They could open shops called Life's Little Pleasures for all they liked. But so long as there were dive-bombing magpies about and snakebite dead in the cemetery, the frontier soul of the place remained intact.

10. AFRICA III

The fish man had just sold Barbara Simpson a beautiful reef lobster. It was speckled grey, rust and cobalt, with delicate feelers longer than its body which it waved gently in the direction of its new owner. Ferdinand brought a bucket of water and gently placed the lobster in it.

"Cost 150 bob," muttered Barbara from her verandah chair.

"Never mind," I trumpeted. "With butter and lemon juice..."

"No, no!" A horrified Ferdinand rushed to take up a position between me and the bucket. "We are returning it to the water." Barbara and Ferdinand loved their reef, and anything that lived upon it. No question, the lobster was going back.

The bemused fish man, meanwhile, had been detained by Jonathan B. at the back of the bungalow, where the staff tended to gather. The fish man, it transpired, doubled as *m'ganga*, or healer, and Jonathan B. wished to consult him. About a week earlier, Jonathan B. had been bitten in his sleep by a centipede. It had nipped him on the foot, and the bite had been troubling him more than he might have expected. It itched like crazy. Last night, he had got his wife to squeeze the bite. Some pus had come out, but the foot was badly swollen this morning.

Jonathan B. was worried. He wondered whether a mchowi might be involved.

"Mchowis resent progress and success," he explained as the m'ganga prodded at his foot. "I have got my shop, I speak good English and am an educated man. I don't know. Perhaps somebody has a grudge against me, and some mchowi is only too happy to assist."

Jonathan B. was ashamed of such thoughts, but the recent death in Kilifi of a young boy from snakebite had put him in a superstitious frame of mind.

"They say he was kicking a crumpled newspaper ball in the street—ssstt hurts!—which came to rest near a plastic bag that concealed a snake. The snake bit the boy. It was a bad bite but after two weeks in the hospital he seemed to have recovered. So they sent him home, only for him to die suddenly and soon afterwards. So I'm wondering whether there's a mchowi about. Perhaps that centipede is going to see me off."

On the Kenya coast, Jonathan B. explained, anything from a centipede bite to a thorn scratch could be transformed by the mchowi's black arts into a fatal snakebite. It happened all the time. People just sickened and died, and everyone knew to blame the mchowis. It was the m'ganga, a benevolent herbalist claiming his share of magic moments, that you called to combat the mchowi in such circumstances. Our m'ganga was called Mr. Guyo Dadi Saidi. He had a fine, wizened face with large, pooled eyes. His hair was a curly fuzz fast turning to grey. He went barefoot and wore mud-coloured shorts that were impressively shredded by age, sun and by contact with fish. Around his belt, he wore a string on which pieces of brown dried bark were threaded. He lived in the fields behind the shop opposite Barbara's plot, and had supplied fish to the local Europeans for years now. As he examined Jonathan B.'s foot, so Barbara's dachshund, Bumble, concentrated on the m'ganga's, licking his toes with relish. Perhaps there was fish there.

Mr. Guyo was shaking his head.

"No mchowi," said Jonathan B. with relief, translating the m'ganga's words. The diagnosis was good. "He does not see the work of a mchowi in this centipede bite. He sees a centipede bite. He also says I am lucky. Few m'gangas can combat a mchowi at the height of his powers. And he is not one of them." The unassuming Mr. Guyo, it emerged, had only begun as a m'ganga six years earlier when he took over from the retiring community m'ganga, the revered Mr. Pokomo.

As we spoke, I could hear drifts of talk reaching us from the front verandah, where Barbara was entertaining friends who had just turned up with the latest copy of the *International Express*. News from England.

"But, my dear Barbara, how *can* you blame Camilla?" came a voice. "I mean, what on earth has she done wrong?"

"Apart from have her claws in him for the best part of his adulthood, you mean? Apart from driving the heir to the throne to adultery?"

"It's what she hasn't done is what she's done," said someone, cryptically. "What a charming lobster."

Through Jonathan B., I asked Mr. Guyo what had made Mr. Pokomo so special.

"He had the knowledge," said Mr. Guyo quietly. "A gift from God. He saw the vision."

"The vision?"

"The tsasapala, the black mamba, biting at the root of the *munyahi* tree."

"Why?" I asked, stupidly.

"Mr. Pokomo was privileged to see the fight, the great fight between the mamba and the puff adder. After receiving several bites, the puff adder died. Mr. Pokomo, he followed the black mamba to the munyahi tree. He watched the black mamba chew at its root before returning to feed the pulp of the root to the puff adder. That snake was miraculously restored to life. Now, I have the gift of healing which Mr. Pokomo passed on to me." I was reminded of Underwood in Australia, and the leaf-eating goanna.

"Why to you? And how did he pass it on?" I asked.

"Mr. Pokomo had been looking for a suitable candidate to

take over the practice, and to initiate into the secrets. Eventually, he selected me. I already had an important position, selling fish, you see. In a special ceremony, I affirmed that Mr. Pokomo was the lord of our community since he had healed many, many people of snakebite. I also gave him a lot of money. And fish. After that, Mr. Pokomo grasped my shoulder with one hand and the munyahi tree with the other, and so his skills and his knowledge of dawa passed to me."

"Of course," said an English voice, "we all know who Barbara's had lunch with."

"Lunch?" a puzzled Barbara replied. "What is that supposed to mean? Ferdinand! The lobster. Please. It's waving at us."

"You know. 1958."

"Nineteen fifty-eight? Oh, you must mean the Queen Mother. When Sandy, my husband, was the D.C. at Kitale. February 13. A Friday. Saddle of beef. I remember it well. We had a corgi and a labrador in those days."

"The Royals have changed, of course," a resigned voice ventured. "Gone down the plughole, they have. England's changed. World's bloody changed."

Becoming a m'ganga had proved a demanding process. There was much to learn. One of Mr. Guyo's first commissions had been blessing the seeding of the crops.

"I would mix the seed with various herbs," said Mr. Guyo, "and plant out the four corners of the field. Then the farmer would pay me and I would leave him to finish off the job. Some of my seedings were very successful. The farmers had excellent harvests. I had many seedings to bless. Now, of course, I know much more than how to bless fields. I cure people who are mad and women who have trouble in pregnancy. I also heal snakebites."

"How do you do that?"

Mr. Guyo looked at me. For a moment, the only sound was

of Bumble licking the m'ganga's toes. I passed over a few crumpled notes, then a few more. Mr. Guyo began to whisper. Jonathan B. strained to hear.

"Mr. Pokomo saw how the bark of the munyahi tree cured the bite of the mamba. I too take the munyahi bark, see how I keep pieces of it upon my belt, and chew it up in my teeth until I have mush which I place inside the victim's mouth. I also smear more mush over the bite. This I do three times. In time, the buried fangs will leave the flesh. With much pus. And the patient will begin to recover." Bark, I recalled, was a traditional snakebite remedy elsewhere. Like Mr. Guyo, white farmers in South Africa had sewn squares of remedial bark into their belts.

I asked Mr. Guyo how effective his treatment was.

"I have never lost a patient," he said proudly. "My patients tend to recover within a few hours. The only thing is that it is important not to let the patient sleep. It does not even matter whether they are people or goats or cattle. I have cured them all. I have even saved chickens. You can ask the people down at Plot 31."

"I remember when I flew to England." Barbara was reminiscing.

"Kenya Airlines or B.A.?" someone asked her politely.

"No, no," said Barbara. "When I flew to England, silly. Nineteen forty-nine. In a two-seater Piper Cruiser. Took eighteen days. I can still remember the stops. We left Nairobi for Wajir, then Moyale, Addis, Djibouti, Asmara, Kassala, Khartoum, Atbara—now in Atbara, I'm sure it was Atbara that..."

What Mr. Guyo, only lately a m'ganga, could not do was inoculate in anticipation of snakebite. This tribal treatment had once been commonplace in East Africa, and among those to have experienced it was the inquisitive Ionides. With a razor blade, one of his men had nicked him on the wrists, elbows, knees and ankles, and applied a mixture of roots, shrubs and castor oil to the cuts. This was an abridged version: the mchowis of the Wakamba people, who live around Machakos south-east of Nairobi, used to slash captured snakes and their own flesh, mixing the bloods before adding it and the snakes' heads to a secret mixture that cooked up into a substance like bitumen and thence to a grey powder that was applied to gashes all over the hands, the feet, the thighs, the buttocks and the shoulders.

Nobody, however, could compete with the Wanyamesi of Tanganyika in this respect. Their unrivalled snake dawa apparently contained the dried heads and tails of boomslangs, puff adders, cobras and mambas; the leg sinews of vultures; the brains of night owl; the eyes and nose of wild dog; the eyes, larynx, tongue and stomach (including the hair balls within) of lion; and, as a final touch, the powdered brain of an insane man. This was reduced to a fat and applied to two hundred small incisions all over the body.

". . . Wadi Haifa, Luxor, Cairo, Mersah Matruh, Benghazi, Tripoli, oh Tripoli!, Gabes, Tunis, Ajaccio, Nice, Lyon, Paris, Le Touquet, then…then we got lost somewhere north of London. We put down in a maize field and asked a labourer where we were. Very polite he was but he didn't understand a word. He turned out to be a refugee worker, from Poland, I think. It was the first time on the whole of that journey that English failed us. Anyway, it eventually transpired we were just outside Melton Mowbray."

"And do you treat all snakebites the same, Mr. Guyo?"

"All except the puff adder," he told me. The m'ganga's reply was not surprising. Of those dangerous snakes that are common on the Kenya coast, the puff adder is the distinctive one, in appearance, habit and in the systemic and symptomatic effects of its poison. "With the puff adder," Mr. Guyo explained, "you take the root of the *mutsulapengo* tree, make it into a paste with the mud of a termite mound and then apply it all over the swollen area."

The look that Mr. Guyo cast in my direction suggested concern for my welfare. I guessed he had begun to assume the mzungu's interest was more than theoretical. I took his next piece of advice personally, for it seemed meant so.

"Watch out for puff adders," he said. "Their bites harm men much more than women. Mambas and cobras, the women feel those worse, but a puff adder could really hurt you. Even with my treatment."

FEBRUARY 1997

Pete Bramwell's second bite was from a puff adder.

"It was—more coffee?—March 1966," said Pete, tamping down his pipe. "Just a few months after the mole viper got me. A collector had turned up at Takaungu. Had a horned viper, *cerastes,* which he'd got in Tunisia. He wanted to do a straight swap for one of my puff adders. That was fine by me—until my snake clobbered me as I was handing it over. Clearly wasn't going to miss out on the very last chance it would ever have of having a go at me. Now, that was painful. The whole of the right-hand side of my body swelled up terribly, and stayed that way for weeks."

The snake park, however, was going from strength to strength. Pete was busier than ever at Takaungu with the snakes, the rats and rabbits. Maintaining the display cases was time-consuming. White ant infestations were a recurring problem; a treatment of diesel sludge from old engine sumps, Pete was gratified to discover, rid the cases of them. Soon, Pete was supplying mambas to antivenom producers. On good days out in the villages, he might bag as many as twelve greens and a couple of blacks. These he placed in bags run up by a tailor in

Mombasa and then packed them in customized tea chests that he put on the Nairobi train. In the evenings, he often gave lectures to the tourists at the Mnarani Club, taking a python and a boxed cobra along. Then there were calls out of the blue from homes along the coast, begging Pete to remove the snake they had just found in the garden, behind the fridge or under the sofa.

Just nine months after the puff adder bite, a few weeks before Christmas, Pete took his third bite. This time, it was from a green mamba.

"It was a bad feeder. I spent so much time trying to get my snakes to feed. I was at the Kilifi shack, trying to force feed the little blighter when it caught me on the thumb."

It was a small snake, maybe two feet long, but Pete knew that snakes didn't come much more venomous than green mambas. He drove straight home to Takaungu where he had a supply of a new mamba antivenom, produced by Behring in Germany. He injected himself in the bite site, and in the upper and lower arms and legs, on both sides of the body.

"Well, after a few groggy days, I felt OK. The first thing I did was go down to Mombasa and buy up the chemist's entire supply of the Behring antivenom. When people heard I'd survived a bite from a green, that Behring juice started selling very well indeed. But what the green mamba bite taught me, more than anything else, was to be calm. No more tearing down to the front of the ferry line."

It was to be just over ten years before Pete got bitten again.

JUNE 1996

Voices from the verandah.

"He's doing what, Barbara?"

"Looking for that chap who got bitten by a black mamba down at Kilifi."

"Ooh err. Rather him than me. We should be going, darling."

"Wasn't that *terrible*? Didn't he die or something? And then, when they put him on a ventilator and brought him back, he was paralyzed. Couldn't move for days. Not a twitch. But he could hear all right. Every bloody word."

"That was it. He heard the doctors talking to his wife about him, didn't he? Telling her that he was most likely brain dead. And there was nothing he could do to tell them his brain was functioning just fine, thank you very much. What a nightmare."

Mr. Guyo rose to his feet.

"Be calm," he said. "That's the best dawa of all. There is a bird called the white-browed coucal that is often found around mambas. We tell a story of how the mamba decided to test the power of its poison one day, and persuaded the coucal to help. The mamba lay in wait and bit a passing man. The man did not see the snake. Instead, the coucal flew up and the man, guessing it was the bird that had pecked him, went happily on his way. When a second man passed by, the snake bit him and then revealed itself. Within seconds, the man had dropped dead."

Mr. Guyo had run out of advice. He took my hand in his, smiled a little seriously at me and disappeared among the cashew trees, Bumble lunging at his toes with a small, pink tongue.

Barbara's guests had left, leaving warm coffee cups on the verandah and a crush of lipstick-smeared menthol cigarette butts in the ashtray. I walked down to the beach. Beyond the dusty casuarina trees I kicked off my shoes to feel the fine white sand between my toes. The sky was colourless. It merged seamlessly with the ocean. Closer to shore, the waves broke in a white line along the reef. Then, where it shelved into the shallows, the water turned ultramarine and a yellow dinghy tugged at its anchor line. Ferdinand was standing in three feet of water. He was holding the lobster just beneath the surface, helping it acclimate. Its feelers moved with the ocean swell.

11. INDIA I

The thing reared up in that menacing shape you're never likely to forget . . .

　　—Gordon Sinclair, *Foot-Loose in India* (1933)

AUGUST 1996

The bus from Madras sped into Mamallapuram. It scattered chickens, children and yellow autorickshaws, and buffed up a roadside cow's flank, a flank shiny from a history of such contacts, before it came to a sudden halt, with a final hoot of its horn, in the town's main square.

It was hot, and the bus smelt of food and animal, diesel, brilliantine and sweat. The passengers and cargo, packed almost to the ceiling, had cohered into a perspiring, bus-shaped mass which was extruded through the door to unpick itself and gradually reshape as individuals and objects. People pulled coconuts from their laps, freed the shopping bags pasted against their armpits and peeled their trouser legs away from those of their neighbours.

I grabbed my sweat-soaked shirt in the region of my nipples and tugged at it. It came clear of my chest with a long, wet sucking sound, like surf retreating down a beach. Heavenwards, where I searched for signs of a forgiving ocean breeze, was only a tangle of black telephone lines and electrical wires, a web of spidery madness spun against a heavy grey sky.

I had not been long in India. To look down the dusty roads that led off the town square on every side, I found, was to suffer a mild kind of vertigo. The meandering dogs and bicycles, the arcades of shops selling plastic cricket bats and hockey sticks, cheap notebooks, cigarettes, baths and telephones. Through a blizzard of stone chips, the pavement artisans hacking at marble, jade and soapstone statuettes of favourite deities from Hinduism's bewildering pantheon: Ganesh, the elephant god of prosperity and fortune who wears a snake for a belt; heroic Krishna; Garuda, the great bird; Nandi, the bull of great Shiva; Hanuman, the mace-wielding monkey; and, in what I took to be an imaginative if now dated attempt to woo British visitors, a life-sized statue of Margaret Thatcher. The posters proclaiming "XXth World's Poultry Congress and Exhibition. Most prestigious poultry event. 1st time in India"; and those of campaigning rationalists which read, "Don't teach your children to worship God. Do teach them swimming."

But not in the stagnant, litter-strewn waters of Mamallapuram's temple tank, the large, square formal pool that traditionally combined the functions of reservoir, washing room and gathering place in the towns of South India. Around the walls of the tank, which had largely fallen into disuse, ads for wood polish and for local beach resorts had been painted. Goats grazed on the flights of stone steps which edged it. The scene was redeemed by flowering pink water lilies, by foundering husks of coconuts suggesting exotic ceremonials, by hovering squadrons of dragonflies, and by a supine *sadhu*, a holy man, who wore nothing but a celestial smile. His skin was mahogany and he bore a sacred trident. From a certain angle, perched on the tank steps below him, lines of casuarina pines seemed to sprout from the pitiful rack of his belly, and march towards the ocean.

Tourists in shorts and rucksacks loitered uncertainly outside shabby hotels and guest houses, and dinky restaurants with names like Luna de Magica sported gingham tablecloths, old candles stuck in the tops of gin bottles and Western music turned down low. Stout Kashmiris stood in the doorways of their jewellery shops, touting for business. The fishing port of

Mamallapuram had recently discovered tourism. But in the distant outline of the town's romantic shore temple, which ocean spray and wind had riddled with holes and crevices over the centuries, was evidence of Mamallapuram's more venerable lineage. Looking inland from the square, I could see part of the town's famous bas-relief, Arjuna's Penance, carved upon the face of a huge brown boulder. I shouldered my bag and walked towards it.

It was a remarkable work of sculpture, an enormous, pulsing tableau, an Indian landscape peopled by figures, apparently part-real, that seemed to be in the process of extricating themselves, over the millennia, from the hold of the smooth rockface in which they had been born. Men, women and children, water-carriers and penitents, monkeys and pigs, elephants and dogs, cats and mice were arrayed on either bank of the great, vertical gorge of the Ganges, Hinduism's sacred mainspring and the composition's central feature.

My eye was drawn, however, to ground-level and the mouth of the river where an impressive, largely lifelike cobra was sculpted, its tiny vigilant head perched atop the oval dish of its raised hood. Immediately above it, in the higher reaches of the ravine, stood a creature half-human and half-snake; a snake's tail merging with a woman's bare-breasted torso. And from the nape of her neck, the hoods of three cobras emerged and were artfully arranged, like petals, behind her beautiful, human head. She was one of India's mystical *nagas*—from *nag*, the classical Indian word for cobra—that are variously regarded as water spirits of precipitation and fertility, and semi-divine acolytes of the great Hindu Trinity.

Writ in ancient stone, these 1,300-year-old representations hinted at the extent of the cobra's symbolic and mythic significance in India. In Arjuna's Penance, the cobra was portrayed both as recognizable living creature and also as the naga, that highly versatile mythic hybrid. Elsewhere among Mamallapuram's rock art, the snake appeared in the form of a giant serpent upon whose coils Vishnu, orderer of the universe, reclined.

In popular Hindu iconography, Shiva, the great destroyer and ultimately the primal life force, is portrayed garlanded with cobras. The cobra also figures large as protector and helper in Buddhism, the religion which once dominated India. A great cobra kept wind and rain, flies and gnats off the Buddha, by enveloping the Blessed One in protective coils as he meditated. When the weather improved, the cobra threw its hood over the Buddha, parasol-like, to keep off the sun. Cobras then demonstrated the extent of their versatility by combining to form a bridge of their hoods so that the Buddha might safely cross the River Ganges.

Over the past months, I had seen how the snake had largely become a simplistic totem of malevolence, the preserve of marginalized mavericks, men of God and medicine men, in the West and elsewhere. But in India, it was as if the whole nation had taken to the country's best-known snake. Here, the blessed cobra stood centre stage, replete with resonance and meaning. Snakes, images and representations of snakes, and snake charmers were ubiquitous. It was as if India had redeemed the snake and invested it with enormous, if complex, spiritual possibilities. In India, you might just learn not to fear the snake.

Of course, the cobra was no mere helping hand to the gods. It often tended to unrestrained vitality; it was, after all, a fertility deity. In some of its myriad mythical manifestations, it was even partial to a little hell-raising. Emerging from the holes in termite mounds that are even today regarded as the entrance to their snaky, subterranean world, the nagas were thought to take to the upper earth to cheat and seduce mortals, if also to inspire them. Their venomous breath could reduce a banyan tree to ashes. They tended to also pollute the places they made their homes, turning sweet rivers into bottomless, birdless pools of darkness that gave off noxious fumes, as if celebrating the virulence of their poison over all things.

In Hindu myth it was Krishna, tiring of the Great Naga's appalling environmental record, who vanquished him after an almighty battle and so left in perpetuity his triumphant footprint—the distinctive pattern more commonly referred to as the spectacle marking—on the cobra's hood. But the munificent Krishna forgave the Naga. He even instructed Garuda, the cobra's sworn enemy, that the great bird, swooping out of the sky, should interpret the mark as a sign not of condemnation but of divine protection, and so abandon his attack.

The manner in which Hinduism thus redeemed the cobra inevitably invites comparison with the treatment the Christian serpent received for its part in Mankind's fall: being cursed "above all cattle, and above every beast of the field," condemned to go upon its belly, to eat dust all its life and to live in eternal enmity with man, so that man shall bruise its head, and the snake his heel.

In India, where venomous snakes are common, officers of the Raj and early Western residents were quick to "bruise the heads" of any snakes they encountered, as their own, distant culture had taught them. They dispatched snakes unceremoniously, regarding them as no more than deadly pests. Local people were regularly shocked by the murder of these sacred creatures, and feared the killers would suffer divine reprisal. J. H. Rivett-Carnac, who lived in India during the late nineteenth century, remembered an Indian he had met in Nagpur. "He worshipped the nag and nothing else," wrote Rivett-Carnac. "He worshipped clay images of the snake, and when he could afford to pay snake-catchers for a look at a live one, he worshipped the living snake,...if he saw a nag on the road, he would worship it, and believed no Hindu would kill a nag or cobra..."

Few of the British in India seemed to feel much kinship for such attitudes, even if they admired the consequent fearlessness of Indians in the presence of snakes. For their part, the British tended to display the sort of no-nonsense sangfroid in the face of snakes that the Empire set such store by, much as their compatriots in Kenya did when confronted by the Tsavo maneating lions during the 1890s.

The other way to deal with snakes, of course, was to demonize them absolutely; to exaggerate the danger they represented to such an extent that a certain fear of them might be considered admissible, even prudent. That way, you deflected accusations of that ultimate imperial disgrace, cowardice, and your letters home no doubt generated extra admiration into the bargain. Such seemed to be the strategy of Gordon Sinclair, a Canadian reporter for the *Toronto Star* who characterized himself as a "news chaser" and "foot-loose prowler" and described the 1930s India in which he travelled as a "he-man land of adventure."

But Sinclair, who was the same age as the century, had a problem projecting the intrepid image he so seemed to wish for himself. He profoundly disliked snakes. Every mention of them caused Sinclair's pith helmet to quiver, rather undermining the "he-man" beneath it, especially since most Indians seemed quite untroubled by the creatures.

Sinclair's account of his 1932 journey, *Foot-Loose in India*, is remarkable for its unintentional portrayal of a mind unhinging due to a fear of snakes. Sinclair piles awesome statistic upon blood-curdling anecdote in what amounts to an insistent appeal to his readership that his fear of snakes is a respectable one. In Calcutta, he takes to bed thinking of them: "A swell lullaby that was; cobras, kraits and vipers, the three worst snakes of all in a land where one man dies of snake-bite every seven minutes day and night." On the train from Benares, now Varanasi, he reminds his readers of this fact: "Every seven minutes, night and day, rain or shine, summer or winter, someone is killed by a snake in India."

In Lucknow, the aghast Sinclair quotes a "casual item" in the

local paper: "Cobras killed three people in Lucknow over
the weekend." To demonstrate just how frightening snakes
are—and perhaps how well he is holding up by comparison—he
talks of a newspaper man of his acquaintance whose nerves
have been shattered by experiences with Indian snakes. "He
hadn't been bitten or even attacked," writes Sinclair, "but he
dreamed snakes and talked snakes. He finally collapsed so completely they had to send him home with a guardian." In time,
Sinclair's unspoken challenge to the readers back home—*and
you're seriously telling me you wouldn't be scared?*—comes
across loud and clear.

In the grounds of a Bombay hotel he comes across a cobra
terrorizing some hockey-playing schoolgirls. He describes a
round of golf near Delhi where the bunkers carry signs reading
"Beware of Snakes," and where it is the assistant caddy's job to
clear the rough of snakes. "In case you happen to be right in
the cobra belt," he writes, "this ball hunting caddy will have a
mongoose on a string and when the ball hits the rough he sends
the snake killer into the long grass." The unfortunate Sinclair
comes across snakes on tennis courts, on the streets and, most
spectacularly, in the bungalow of a small provincial town north
of Delhi; showdown time.

> With the first peep of dawn I was up to see where on earth I
> was…As I moved the tin bath tub, the head of a brown cobra
> came around the edge with a menacing sweep and I jumped
> straight in the air just as he pounced. He hit the side of the
> tub and it thudded like a drum. I was over the top of him
> marking myself down as a fool and an idiot. Here with a
> drain into the bungalow fairly coaxing snakes to come in for
> drink and shade, I was snooping around with loose slippers
> and bare legs. I slapped the bathroom door shut and got into
> jungle clothes—high boots with leather breeches. Then I
> picked up the pistol and the whippy cane…I let go three
> shots and he went sliding back in the corner like a long rope

of spluttering sausages...I let him have three more and that was the end of my first cobra. I went up beside him and he got the seventh bullet straight through the head.

Sinclair, I might have said, you just put seven bullets into a cornered snake, a modest reptile, spluttering sausages, not a frenzied rogue elephant. Until, that is, I remembered my own cowardice and saw myself lining up alongside this peculiar Canadian, emptying the barrel of my own pistol into the same, small creature, just for good measure.

The rather more intriguing notion—that Sinclair had just blown away a Hindu divinity—neither detained nor distracted our hero. Sinclair clearly saw himself as more of a roving reptilian white hunter than a comparative theologian. Sinclair dismissed Hinduism as no more than "one of those superstitious outcroppings of black magic which makes this pagan peninsula the wildest east of Suez."

At least, Sinclair did not condemn the Hindu attitude towards snakes as devil worship. He left that to a series of indignant British clerics including the Reverend John Bathurst Deane who wrote in 1830 how "it was [Satan's] device, therefore, that since by the temptation of the serpent man fell, by the adoration of the serpent he should continue to fall." "All over India," the Reverend W. S. Durham wrote in 1948, could be seen

> the little platform built over a snake hole, with its pot or wicker basket and vermilion representation of the cobra; or a crudely hewn granite block with a rough-carved cobra smeared with vermilion...how absolutely typical of the devil's tactics was the way in which he encompassed man's doom by hellish subtlety, and then induced man to worship him for the very subtlety which destroyed him...Could anything be more tragically eloquent of the spiritual bankruptcy of our age, when the devil appearing as an angel of light is welcomed as God Himself?

Even today, we had stopped near one such termite mound "shrine" which stood by the roadside on the way to Mamallapuram. It had been "founded," it could be presumed, by a Hindu devout who had happened upon a cobra entering or leaving the termite mound—revered as the snake's favoured home and, by extension, as entrance to the snake's mystical underworld, place of groves and lakes where the lotus grows, of fragrant perfumes, warm sun and the music of pipe, lute and tabor. The first traditional offerings would soon have appeared around the entrance, perhaps a scattering of parched rice, turmeric and some *ghee*. Then dried flowers and mint, a little sandalwood, saffron or betel leaves would have been left.

And so the shrine had established itself. A thatched roof had been constructed over the mound from which a portrait of Shiva had been hung. At the front of the tomb, a holy trident had been set vertically in a concrete plinth, and in an Elf Axle Grease tin, hung up to receive donations, a small coin slot had been cut. There was also an arrangement of upright stones perhaps two feet high daubed with red and yellow spices, Durham's blocks of "crudely hewn granite," on which cobras had been carved in a variety of postures, and intertwined with lotus leaves. In time, walls might be erected around the shrine. One day, a Brahmin priest in saffron robes might appoint himself its caretaker, a car park might even be created and, by and by, to W. S. Durham's disgust, the shrine would grow into a temple.

W. S. Durham claimed devil worship, but reached that conclusion by failing to appreciate that Hindus were worshipping something far removed from the fundamentally evil, Christian version of the snake. The snake's Christian criminal record, the tenacious link between sin and the serpent, has never been recognized by Hindus, who have bestowed very different associations upon it, particularly upon the cobra. In India, the charismatic cobra is the totem of new life. Like other snakes, cobras tend to appear with the seasonal rains, which modern herpetology puts down to rising flood waters and so sees snakes as mere refugees from the wet. In India, however, they remain widely regarded as harbingers of the harvest, and suggestively shaped harbingers at that. The cobra presides over new life in the broadest sense. Specifically, it is an unabashedly phallic representation of the sort which, one is tempted to assume, would have tipped Reverend Durham into apoplexy.

"There are certain obvious points," wrote Rivett-Carnac, cutting as near to the chase as he felt he comfortably could in the mid-nineteenth century, "connected with the position assumed by the cobra when *excited* [my italics], and the expansion of the hood, which suggest the reason for this snake in particular being adopted as a representation of the phallus and an emblem of Shiva." Most temples dedicated to Shiva have a lingam, a phallic emblem which unwary Europeans—or W. S. Durham—might just mistake, or choose to mistake, for a bollard, centred at the heart of the shrine. Many of these, to emphasize the sexual connection, have coiled cobras carved around them.

It was when I saw a woman in a sari standing before more of those strange upright cobra stones, a little group of five arranged around a statue to smiling, prosperous Ganesh under a shady sacred *pipal* tree beside Mamallapuram's temple tank that I finally realized what they were: the offerings of childless local women to the gods of procreation, the *nagakals* or snake stones which can be found clustered by shrines and under holy trees all over South India.

From a steel dispenser, a shallow tray on which piles of coconut husk, flower petals and *garam*, bowls of water and of ghee were arranged, the woman sprinkled offerings upon one of the stones and fingerprinted it with daubs of yellow turmeric and a purple ceremonial powder called *gulal*. The stone, about the size of a headstone, commemorated the woman's greatest wish: offspring. It was a graveyard not to the dead but to the unborn, pleas to the fertility deities to grant children. As I turned away, the wind lifted the petals and set them gently down upon the waters of the tank.

The afternoon was drawing on. I found a room at the Lakshmi Lodge, a whitewashed concrete building arranged around an interior courtyard down by the beach. The Lakshmi stood alongside an abandoned brothel. As if to scotch the association, the walls of the Lakshmi were plastered with distractive sylvan scenes—fishermen, rivers, swans and fireball sunsets—and accompanying platitudes. "Oh, how beautiful is love," proclaimed one. "And how beautiful life is for those who share it." Not that the Lakshmi staff didn't have a business to run. "After 11pm, restrant close," barked a gruff adjacent notice. "If you want speak, go roof top." It was only sunset, but I grabbed a beer and went roof top anyway. From here, with the low sun flaring at my back, I could look out along the Coromandel coast where the last of the postcard sellers prowled the beach for lingering tourists, and reflect on why I had come to India.

After America, Australia and Africa, the Indian attitude towards the cobra seemed uniquely positive. It recalled the cobra's status in ancient Egypt where it was the honoured emblem of royalty, and the place of the snake in the Greco-Roman world, where it was the symbol of healing. Indeed, many pharmacies and medical organizations such as the British Medical Association continued to include the snake in their logos. I suppose I had the idea that exposing myself to such an attitude might erode my fear of snakes. If I could bring myself to believe in the inherent benevolence of snakes, then what would there be to fear?

I did not see the man until he stood directly below me. He carried a round, flat reed basket and was observing me quietly. He finally greeted me with an expansive wave of the arm. Then, at the last moment, he transformed the gesture, extending the arm towards me in a rolling, sinuous manner that could only indicate one thing. And I knew what lay in the basket.

12. AMERICA IV

"We were saying attempted murder," recalled the man from the Scottsboro D.A.'s office, smoothing his tie down his belly. "We offered him just seven years for a guilty plea. Which meant he'd have been out in even less." He had stumbled upon me among the records of the Jackson County Courthouse, a dusty, ill-lit backroom whose walls were lined with filing cabinets and ceiling-high industrial shelving, where I was working my way through the Summerford trial transcript.

"But he wasn't having it," continued the D.A.'s man. "No sir. He held out for an acquittal. Which, seeing as he had already collected two previous convictions in the state and what with our Habitual Felony Offender Act, was a high stakes strategy."

I nodded knowingly. By now, I'd seen enough of the law shows on TV to nod knowingly. Living out of motel rooms, where TV offered about the only distraction going—once you'd slung the half-eaten boxed pizza in the direction of the bathroom unit, that is, and fended off the uninvited phone-calls offering interesting services of an intimate nature and then locked the door in case the caller thought to come by in person—you got round to catching the law shows. Which is where I'd heard a version of the D.A.'s man's spiel before. If he and all those episodes of *Law and Order* or *Murder One* that he echoed were anything to go by, American justice was about strategy, pure and simple. What mattered, rather than absolute notions of guilt or innocence, was how you worked the judicial system. Turning procedural niceties to your advantage. Getting the best deal you could.

I'd left my Scottsboro motel room that morning. The fancy
wood and glass offices of the *Sentinel* on Veterans Drive were
stippled in the blossom of low-slung cherry and peach trees.
Above the picket-fences and mail-boxes that belonged to the
weatherboard homes you usually saw there, only brief scraps
of door, window and roof were visible. Even the vermilion-
coloured cardinal birds seemed eclipsed by this show of spring,
and rummaged sulkily along the verges. On Broad Street, an arcade of turn-of-the-century brick facades, billows of winter dust
were emerging through the doors that shopkeepers had wedged
to let the sun in. Men in baseball caps emerged from Payne's,
the town's old soda bar, clutching Styrofoam coffees and
doughnuts in paper bags which they clenched between their
teeth as they tugged at the doors of their pick-ups.

I parked outside the splendid neo-classical, red-brick courthouse. It sat among tended lawns in the town square, dwarfing
its surroundings. With its domed, Italianate clock tower and
imposingly columned portico, it was less a structure than a
statement, one that wished it known how seriously Scottsboro
took its judicial responsibilities.

So Scottsboro had the damnedest big courthouse, and no
shortage of freshly painted attorneys-at-law offices scattered
around the square, with flourishing pot plants and modern
blinds in the windows. But it wasn't white-shoe Wall Street. You
could tell that by the courthouse records office which doubled
as the employee's pantry, with a steel sink, a kettle and a small
gas oven where a regular procession of secretaries heated up
sour-smelling lunches from coffee time onwards. Over the
years, those meals had permeated the documents; as I turned
them, the whiff of what I took to be old tuna bake rose from the
pages of Darlene Collins's testimony.

Then there was the man from the D.A.'s office, florid, overweight and wearing a swirly-patterned polyester tie that was
tasteless, and not even cool, Harvey-Keitel tasteless. *L.A. Law*

he wasn't. As the tuna-scented transcript indicated, the Summerford trial had been big time for Scottsboro's legal community. "You will see and hear about things you've never seen before," read the defense attorney's breathless introductory statement to the jurors. "You will never forget this case for the rest of your life." Nor, you could take it, would the defense attorney. From his own words, you could tell he was just as excited about his day in the limelight. It wasn't so often the national media packed out the Scottsboro courthouse and hung on his every word.

The man from the D.A.'s office shrugged. "Summerford should have taken the offer. Of course he should. What I would say is this: take a good offer if ever one presents itself. Meanwhile, the wife's not in the phone book, you say." He chuckled knowingly; the omission did not surprise him. "Let me see if I can find you something on her whereabouts."

The smell of warming lunch was overpowering now. It was a leftovers stew, I think, a meal which had been around the block a few times too many that finally drove me from the building and into the fresh spring air. I walked across the grass and found myself on the far side of the square. Here, the town soon turned shabby. Among the thrift shops and empty lots, a building stood, compact but derelict. Waves of damp had stained the red-brick walls of its open porch, and left white, granular mildewed tide-lines there. In the tall, narrow windows, the iron grilles had rusted and the panes broken. Along the crumbling lintel the words JACKSON COUNTTY JAIL were engraved.

It was a jailhouse from another time. It had a homey, kitchen-style front door, all white-painted wood and panes of glass from half height. On its flat roof sat a mockingbird that opened its beak to sing, then seemed to think better of it as the wind snared along its black and white plumage. The wide porch had the feeling of American legend about it; the kind where resolute sheriffs with strong jaws sat on warm nights, reasoned

with armed men that they knew by daylight as honest neighbours, and flicked glowing cigarette butts into the night to the
rhythm of cicadas.

Which was what had happened in Scottsboro back in the
spring of 1931, when a lynch mob had gathered here, demanding vengeance and eyeing nearby trees for suitable boughs. And
so began an episode that brought shame on Scottsboro and left
the town with something of a record. For miscarriages of justice, few towns were more notorious than Scottsboro. There
were even those who said that prosecuting the law in Scottsboro was like practicing viniculture in Alaska or promoting
women's rights in Riyadh: there were far better places to do it.

It had begun one night in March 1931, when nine black
youths heading south through Jackson County on the freight
train from Chattanooga had got into trouble. There'd been
brawling in which some white boys got their pride hurt. By the
time the police reached the scene, two white females were accusing the blacks of rape, a capital offense at the time.

The oldest of the nine was just twenty when the Scottsboro
Boys, as they came to be known, were brought to trial in the
town on April 6. They pleaded innocent but by the day's end,
the jury had found all nine guilty of rape. When the verdict was
returned, the crowds outside the courthouse broke into ap-
plause. That night, the National Guard was on hand to keep
the crowd from the jail where the Scottsboro Boys were being
held, and a brass band paraded around the square until the
early hours.

There were no offers on the table that night, no deals, unless
you counted being strung up from the nearest tree as preferable
to what the state had in mind. To the incensed mob, the stops
the freight train made through Jackson County—Fackler,
Hollywood, Scottsboro, Larkinsville, Lim Rock, Woodville and
Paint Rock—were places they had known since childhood, familiar and nostalgic. Places where family and friends lived and

farmed. Where they had been to church. Sat down at table.
Now, they would always be known as the places where nine
black men paused from raping two defenseless white girls to
muffle their cries until the freight train carrying them rumbled
on into the night, where nobody would hear them scream. It
was as if Jackson County itself had been violated.

Many in the South took the convictions to confirm their
worst suspicions regarding blacks. The sentences were a vindication of their segregationist, supremacist position. A wide body
of Southern opinion, including many of the newspapers, supported the verdicts. Further afield, however, they were roundly
condemned as unsafe, based as they were on the sole evidence
of the alleged victims. There were no independent witnesses,
and neither the physician who had examined the two women
nor any of the freight train staff were called to testify. In May,
there were demonstrations in New York. High profile figures including Albert Einstein and Thomas Mann joined the chorus of
protest which reached new heights in July of that year, when
eight of the Scottsboro Boys were sentenced to die by electro-
cution.

Several years later, when international pressure finally
caused the case to be reopened, one of the women retracted
her allegation of rape against all nine men. She also disputed
the unchanged evidence of the other plaintiff. Incredibly, a new
jury once more returned a guilty verdict. By the time the
Supreme Court eventually overturned the convictions, and it
was revealed the women had had sex, willingly, with several
white boys on the train that same night, some of the Scottsboro
Boys had spent fifteen years in prison, many of them at the
business end of Death Row.

No wonder, then, that Scottsboro had given its courthouse a
major refurbishing in the years following the second World
War. It was a way of closing a shameful chapter, and making a
fresh start. The problem was that the Scottsboro Boys was a

pretty durable tag, and the only way round that was Scottsboro renaming itself. The tag's superficial jauntiness—they could have been a Fifties school band with hair licks—only served to highlight the infamy of the episode. It was a label that vividly evoked the dark ages of the South. The segregationist signposts. The men in the hoods at night, arranged as orderly as a school photograph, only behind a burning cross somewhere deep in the woods. The rattlesnakes civil rights workers found left in their cars. The whites who gave themselves names like the Cottonmouth Moccasin Gang and killed blacks for sport.

Even today, black churches burned across the South. There'd been cross-burnings in Dekalb, the neighbouring county, as recently as the 1980s. Bath towels with the letters KKK painted upon them were slung onto black porches. Just across the state line in Rome, Georgia, the sight of a black boy and a white girl petting on a school bus had caused klansmen to march on their school. Shots got fired into the boy's home.

So it was that visitors thinking to stop off at Scottsboro for the town's renowned market days, on the first Monday in every month, soon found something nagging in the memory.

Scottsboro? Ain't that the place that put nine innocent black guys on Death Row?

Yeah. But these days they too busy handling rattlesnakes for that kind of thing. Heh. Heh.

That, then, was Scottsboro done for. Although sixty years separated them, two sensational trials—subjects: rape and racism, rattlesnakes and religion respectively—in a town the size of Scottsboro suggested a compellingly alliterative recurrence of the extreme themes that were staples of the Southern stereotype. Scottsboro didn't necessarily *look* like the home of lynch mobs and preachers who set about their wives with rattlesnakes; it simply and demonstratively *was*.

The man from the D.A.'s office had news of Darlene.

"People say as she went back to the Mountain," he told me.

"Back to Sand Mountain, where she came from. Got herself a trailer home somewhere near Dutton. You could ask at the store there. But you watch your feet. They're different the folks up there."

~

On the Saturday morning, Glenn took Darlene to the stores. She had woken early when the light off the soybean fields flooded through the windows to find she'd rolled off the couch during the night. She had vomited. That she had somehow slept through that, and the pain, she could hardly credit. Her arm had swollen hard as a tire. The thumb was turning black, and vivid streaks of red like night-time fireworks were fizzing down her wrist. She had never felt so bad. She thought she might die that very moment, on the floor, in a puddle of her own vomit. A few minutes later, she heard the sound of a car. It was Glenn.

"I just brought the Chevy up from the woods," he told her. "You need the hospital."

For a moment, Darlene thought it was all over.

"But first," Glenn added, "we got a few errands to run."

It was odd, thought Darlene, Glenn running errands even as she was dying from a rattlesnake bite. God, she wanted the pain to go away, she wanted to get to the hospital but then she was in no condition to do much about it. Her husband won arguments even when she was at her best. Now was not the time for straightening him out. So she did not complain as he drove to Video Movie World to drop off some overdue videos. She even found herself wondering, as he got out of the car and she caught a glimpse of the titles, *Malibu Hot Summer*, *Wraith* and *Kandyland*, which of those movies she'd seen and what they had been about. At the ABC Liquor Store Glenn picked up a bottle of vodka. At Finamart, just off the interstate, he loaded up with Winstons, cola, orange juice and microwave sandwiches. Then he turned onto the interstate heading north.

"That's not the way to the hospital," said Darlene. "You said you'd take me to the hospital."

But Glenn had a plan, and the hospital wasn't part of it. Kill her by snakebite, but make it look like suicide. Like Cleopatra, the most famous suicide ever. The Chevy came off at the next exit, took Veterans Drive away from town, turned left into North Crawford and took a right. They were back on Barbee Lane.

~

I made for Dutton that afternoon. And the man from the D.A.'s office was right; something was different about Sand Mountain. Not that it was much of a mountain, was Sand Mountain. It rose, what, a few hundred feet from the Tennessee River valley. So it wasn't the sort of mountain you could trek up, or ski, or conquer. You couldn't plant your flag upon it. You couldn't fly into it. What Sand Mountain did was keep Scottsboro's residents from the early morning sun; in the winter, this could mean the frosts clinging on as late as midday. Sand Mountain pushed the transmissions of passing cars into low gear. Might even, on a hot August day, cause the odd, aged pick-up's radiator to overheat. When nobody was about, young men returning from Fort Payne liked to cut their engines at the top of the mountain road and see how far the slope would take them, the Tennessee River bridge being good going. Sure, Sand Mountain was steep. But only as long as it lasted.

It even looked the same. The cornfields, churches and chicken farms all capillaried by a mazy network of country roads that converged on crossroad townships boasting the same few amenities: the tanning studio, the gas station, the video rentals painted an invariable pink and a general store humming with talk of soybean prices.

Still, something changed on the way up to Sand Mountain. It was as the road out of Scottsboro began climbing, and wound its way deep into sycamores, beeches and tangled briars, that the sunshine was reduced to a dapple before the foliage crowded in and the light was obscured altogether, if only for a moment. When the car re-emerged into the daylight it was as if an indefinable border had been crossed and I had been ejected into another world. I was somewhere else. I understood why they called it a mountain. There might not be much altitude to brag about, but Sand Mountain felt like a place apart.

It had always been this way. Up here, you were a distance from the bustle of the towns that had grown up in the valleys, on the river fords. Sand Mountain people had settled on the margins. Modern America made it here in bits and pieces, in the shape of sunbeds and satellite dishes, but had never truly claimed the place. Up here, change had a tough time of it. The man from the D.A.'s office, I remembered, had told me how Darlene had gone back up to the mountain. He had made it sound less like a geographical relocation than a whole social and emotional repositioning. Which was understandable, after what Darlene had been through.

In ten minutes, I was in Dutton. The settlement was islanded among long fields that lapped at it like a threatening high tide. There was every prospect, I mused, that a day-dreaming farmer might one day plough the entire place under and, except for the odd bump, not notice till they got round to harvesting the corn, and found scraps of weatherboard, bedding, tarmac and school books, the till from the general stores and the odd broken video cassette among the cobs.

I pushed open the door of the general store to the sight of a group of locals in overalls who had formed a tight circle around a man, and were gesticulating instructively at him. One held his hands in front of him, splayed his fingers and twisted them against each other. Another spiralled his arm upwards until an unseen obstruction stopped him at the elbow. For a moment, I sensed I'd stumbled across some backcountry ritual that I was not supposed to witness. It turned out that it was a collective

bid to convince the man that he was holding the right part for
his tractor. The man examined the part for a long time. He
turned it over, as if doing so would eventually expose some reassuring feature that had so far escaped his notice. As he did
so, you could hear all Dutton holding its breath; *would he take
the part?* Then the man spun it in the air and caught it neatly
on the descent. He was decided.

"You say so, Bud," he said, slapping Bud across the shoulders
so Bud smiled and Dutton smiled with him. "Though sounds to
me like you've got mighty strange ideas about my tractor."

The posse were still talking tractors when I asked the store-
keeper for the way to Darlene Collins's place. His eyes narrowed with disapproval. What was unclear was whether it was
me, the nosy stranger, or Darlene with all the trouble she'd got
herself in, that he didn't take kindly to. So I took the directions
and left smartly.

She lived in a white, mildewed trailer home on the edge of a
cornfield, one of a dozen trailers arranged around a dirt-road
junction. It made for a shabby settlement. Abandoned bikes
protruded from the mud. An umbrella-shaped clothes line had
fallen to earth. Then, with evident relish, somebody had tram-
pled upon it. One stretch of line ran, improbably horizontal, a
few inches clear of the earth so the clothes pegs that still hung
there tinkled gently in the breeze. There were cars, mostly car-
casses now. Engine parts, doors, wheels and windshields were
missing, snatched to serve other cars. Bits of blue sky were
wanly reflected in the puddles of recent rain that littered the
ploughed fields. Garbage was piled high in a track-side skip.
The smell of putrefaction was vaguely medicinal. Behind the
trailers was a dark run of woodland to whose trunks No Trespass signs had been nailed. The men who stood around in
sleeveless T-shirts and bandanas, tattoos on their arms and
crushed beer cans at their feet, lent the signs any authority they
might have lacked. Their narrow eyes tracked my stranger's

car into somebody else's line of vision. To the inside of several trailer windows, the Stars and Bars of the Confederacy had been nailed.

Darlene's trailer lay on the edge of the settlement. Out front, odd bits of plastic piping lay around a gleaming satellite dish. I instantly recognized the woman who answered the door from the press photos: the thick ginger hair, the freckles against a milk-white skin, the eyes, the strong jaw. She was not so much unwelcoming as wary. She'd had trouble with journalists, she told me. Some of the things they'd called her. Anyways, I was welcome to join her in a beer, she said, but I'd have to excuse the mess.

I stepped inside. A black plastic dustbin stood in the corner of the kitchen, piled high with crumpled Budweiser cans and junk-food packaging. A slagheap of dishes loomed above the sink. Darlene handed me a beer and motioned towards the other room where two people were slumped before a large TV screen, piled ashtrays perched on their chair arms. That was Debbie, Darlene's sister. Debbie had feral eyes. Had wrecked a car recently when driving drunk. Broke her wrist in the accident. And the young man in dungarees was Chris. Chris looked good-natured and chewed tobacco. He took pains to describe how the tobacco quid moulded itself to the base of his teeth, and seeped brown and nice down his throat. Chris was eight months shy of the state's legal drinking age.

"And Chris," said Debbie, "is just counting the minutes."

I asked Darlene how she was doing.

"I still have nightmares," she said, lighting up another cigarette. "About snakes and lizards. Why lizards I don't rightly know." Her Bible lay on top of the TV, but Darlene had stopped going to church. She'd had it with preachers. These days, dancing down in Scottsboro was more her style. They went pretty regularly, Debbie and her. That's how she'd done *her* wrist, getting in a fight. They needed each other, the two sisters, and never more so than when they went driving. The one would steer with her good hand while the other did the indicating and wipers and whatever with hers.

There was a time after it happened, said Darlene, when all the attention was looking to take her places. They even asked her on *Sally Jessy Raphael*. That was then, though. Lately, she'd been working at some hosiery factory near Fort Payne that paid a useful $11 an hour. She needed to get herself working again and kept telling herself she'd start looking around on Monday. Meantime, she wasn't up to caring for the kids. Marty and the girl who'd been born late the year it had all happened, they were staying with family until she was up to looking after them.

It seemed strange, Darlene in these surroundings. Darlene's story, dark and tainted though it was, had always had a Gothic glamour to it. Snakebite, I'd assumed, would somehow have changed her, not just by the acquisition of experience, but in the form of a more profound transformation, a kind of rebirth, elevating her above the banalities that now confronted her: the drink, the squalor, the fights, the kids she couldn't look after. Instead, she was just a woman seriously behind with the housework. A woman who wasn't coping. Living on the edge of an Alabama field in a trailer that wasn't going anywhere. A woman who needed to find some work and straighten herself out real quick. Snakebite, I now realized, happened in the real world and, once the pain had passed, left the victim in the same place. In snakebite there was no dark beauty, no hidden meanings. It didn't offer salvation. Just inflicted pain.

"From that second bite," said Darlene, "my arm swelled up to fourteen inches. People tell me having a baby's the worst pain there is. Well, I can tell them different."

I asked her about that Saturday morning, the morning Glenn decided against taking her to the hospital. Darlene closed her eyes, hurt by the memory.

Glenn had got her back inside the house, she told me.

Closed the curtains and started drinking. Darlene lay in the half-light, drifting in and out of consciousness, the pain hammering at her. Around lunchtime, they heard a car approaching.

She told me how Glenn told her to keep her mouth shut. He was holding the gun. Steps approached the door.

"Glenn?" came a concerned voice, rattling the locked front door. It was his sister. "Glenn, you and Darlene in there?" For a while, they could hear her padding around outside. Darlene was aware of her every footstep, but she wasn't up to doing anything about it. She heard Glenn's sister turn on her heels and then hesitate, as if expecting a belated greeting to bring her back. Darlene might have cried out then, she told me, at the last moment, but Glenn had sensed the temptation and was aiming the gun right at her face. The car engine turned over, reversed down the track and faded into the distance. Glenn pushed back the curtain to check they were alone, then slugged himself a vodka.

She told me how, throughout that Saturday afternoon in October, she lay up on the couch, whimpering in pain. It was all she could do to ride the waves of agony. All her energies were concentrated upon them. It was as if her hand were under repeated hammer blows, and hot to the point that she could smell it, like the first aromas of cooking rising from a grill. She had never smelt her hand before. Then there were the arrowing jolts all up her arm, and a terrible, queasy sickness that left a taste in her mouth of things she was sure she had never eaten; metal and straw and barley sugar. She only wanted the pain to stop. That seemed simple enough. But there was only Glenn to ask. And Glenn was slumped in a chair in front of the television, drinking vodka. Above the sound of the television, where voices were shouting between bouts of incidental music, Darlene was sure she could hear the distant drone of the interstate. The interstate that ran north, she knew, to Chattanooga. After that, she wasn't really sure where it went.

At the end of the afternoon, Glenn looked at his watch. From where she lay, she could see him raise his wrist close to his face. That way, he could focus on the watch face.

"One hour," he said, returning his gaze to the TV. "You got one hour, Darlene. Then's when I'm taking you back to the snake shed." It was the first thing he had said in hours. And only yesterday, Darlene mused, detached now, God had spared her.

Six o'clock passed uneventfully, she told me. Drink had extended Glenn's deadline. But Darlene was too befuddled by pain to notice. At about seven-thirty, the dusk brought Glenn to his feet. He blinked twice, and dragged his wife outside. Darlene whimpered with the pain but she did not struggle. Her balance was playing strange tricks on her. It was all she could do to remain upright by planting her feet wide. She was beyond knowing where she was until she caught that smell again, the mustiness of old wood and something else as well, and that brought her to her senses and reminded her of his words. The door of the shed closed behind them. In the gloom, she saw that Glenn had picked something up; it was a piece of black plastic water piping about a yard long. He walked Darlene towards a box, the one that housed the rattlesnake she'd handled without incident just the previous evening. He ran the piping across the mesh lid as if he were playing a xylophone. Instantly, the two snakes within rattled long and hard, heavy with menace. Glenn repeated the trick and the snakes' instant response, as if at his behest, seemed to please him. The snakes would do his bidding.

"The lid," he said, massaging the gun at Darlene's temple. "Lift the lid."

She lifted it as she'd lifted it before, but this time Glenn poked at the snakes with the piping, like he was stoking a fire. For good measure. The snakes struck at the piping, then withdrew, their rattles sounding furiously.

"One thing's for sure," said Glenn. "You won't handle that snake no more without it biting you."

This time, he was keeping the Lord out of it.

Darlene shuddered at the awakening memories. She leaned forward in her chair and stubbed out her cigarette.

"I don't know if I can go on with this," she said. There was a sudden noise in the kitchen which made us jump, for we were deep in Darlene's story, and it was as if Glenn had returned to finish the job on his wife. Chris, tumbling the quid of tobacco round his mouth, got up to investigate and shambled into the kitchen.

"Just your trash gone over," he said.

13. AUSTRALIA IV

OCTOBER 1996

I left Port Douglas in the morning, stopping for gas on the edge of town.

"I've never seen snow," said the girl on the till. I put her wistful observation down to the heat. That or my temperate-zone accent which perhaps evoked falling leaves, log fires, pale sun flaring along ice-bound stretches of riverbank and flakes flurrying around street lights.

"No, that's not quite true," she added. "I saw it once, on a mountain far away in Victoria. But my mum had a snow phobia, and wouldn't take us near it."

"It's white," I informed her, relishing the role I suddenly seemed to have been fitted with; a traveller from distant lands with entrancing tales to relate. "It's cold and strangely…"

"Yeah, yeah," she interrupted me. "I have seen it on TV, you know."

As I drove north, beads of sweat were budding at the back of my neck. Each slight shift in my seat sent them coursing downwards to leave licks of salt between my shoulderblades. The leaden sky was littered with scraps of sunlit turquoise, windows in the cloud where the vapour trails of aircraft showed. Just before Mossman, I left the coast road which would soon dwindle to a dirt track as it headed into the emptiness of Cape York, the 500-mile expanse of scrub and swamp stretching to the top of Australia, and drove inland.

I passed through paddocks of a bleached-green crop that I did not immediately recognize. It carpeted the coastal plain, then gave way to rainforest at the foot of distant, dark hills. It grew tightly packed in the manner of bamboo and stood tall, as high as ten feet, culminating in an extravagant topknot of spear-shaped fronds. Along either side of the road, it presented an impenetrable stockade. Then, with a prickle of apprehension, I realized I was in cane fields. With my free hand, I felt for the ash wood charm in my pocket. I was in taipan country.

It was said that sugar cane was to taipans what bloodstained water was to sharks, ice was to polar bears or eucalypts were to koalas. As the taipan became infamous in the 1950s, farmers remarked how they'd always known the taipan, as they now understood the snake was called, as the brown cane snake, for the simple reasons that it was brown and it lived in the cane. Fatal bites received in the cane were recalled by press and public alike. John Pringle's bite near Port Douglas in 1929 had occurred in a cane field when his horse-drawn plough had dug up a taipan. An Italian settler died in 1931 from a taipan bite which, he told his brother, he had disturbed in the growing cane where he had gone to cut tops to feed his horses. The renowned snake collector Eric Worrell even coined the tag "killer of the cane fields" for the taipan. And it was widely believed that burning the cane prior to harvesting, a popular practice in the 1950s, was adopted as a means of ridding the crop of taipuns.

In the public imagination, the cane was infested with taipans. The truth was rather more complex. Combatting leptospirosis, a debilitating disease carried in rats' urine that was communicable through the numerous cuts and grazes that cane-cutting invariably entailed, was the primary reason for burning the cane. Taipans were indeed attracted to cane fields, not least by the numbers of rats there, but they also frequented other habitats including heath and grazing land. Neither was cane their preserve to the exclusion of other snake species. A Mossman cane farmer once reported killing forty-two snakes in the course of clearing a four-acre cane paddock, but they were all death adders.

Despite these discrepancies, the snake's association with cane became an established mainstay of the burgeoning taipan myth, much to the despair of the sugar trade. For when the new settlers, most from Britain and Italy, heard loose talk that the most likely prospect for cane workers was of dying from snakebite, they soon found themselves drawn to alternative vocations. Queensland's biggest export industry began struggling to recruit enough workers. It was facing the problem that Indian tea plantations habitually suffered, when king cobra sightings cleared the hillsides of tea pickers. It was said that the eighteenth-century surveyors along the Virginia–North Carolina border stopped work when rattlesnakes appeared in the spring, and that turn-of-the-century workers in the Melbourne area had threatened to down tools when snakes were sighted in the breweries' barley crop. The threat to the industry was a serious one.

In 1956, the alarmed Queensland Cane Growers' Council went on the offensive. It ran an article called *The Taipan— Truth and Treatment* in its monthly review. "Death by snakebite in Queensland is a rare occurrence," the piece began, "yet continued publicity has done much to produce unnecessary alarm and fear…" "Too Much Talk of Taipans," read a headline in the *Cairns Post* in July of that year, pursuing the same theme. "A great fear of these snakes on the part of many new Australians brought here to work in the cane fields is particularly noticeable," the paper quoted the secretary of the Cairns District Cane Growers' Executive. He also reported how many farmers "had complained about a lack of hygiene in farm barracks"; the men, it appeared, "would not walk at night to conveniences because of their fear of snakes." Branding the region as dangerous snake country, he concluded, would not encourage new settlers.

It was harvest time. Miniature locomotives were trundling along the network of narrow-gauge tracks that connected the cane fields with the roadway sidings, hauling long trains of mesh-sided trolleys piled high with freshly cut cane. At the sidings, where the wagons were emptied into the eighteen-wheeler trucks, the road trains bound for the sugar mills at Mossman, the smell was sweet but soured by occasional whiffs of a turpentine rankness. I pulled up beside a line of loaded wagons. The ruthless harvesting machine had chopped the cane into stringy, two-foot batons. I wondered whether any snakes might have been caught up in the process, whether my first taipan in Australia might be similarly sectioned up and scattered among the loaded cane crop. But there was no such snake. I returned to the car.

I was in prime taipan territory, but I could see no taipans. The only ones I'd so far seen had been back at London Zoo. All I had encountered were references to taipans: in the paper, the seasonal reminder of the danger from snakes and, rather more obliquely, a public announcement that warned of the danger from fallen power lines which were represented as striking snakes; the passing mention, overheard, of a taipan seen on the road at dawn; the headstones of Kevin Budden and John Pringle; a town football team further south called the Taipans; and that snake decoder, misplaced among women's magazines in a Port Douglas newsstand. Up on the Cape, I'd heard you might see the odd snake's head contained in screwtop grocery jars labelled "taipan" on display in the remote roadhouses which served food and fuel. Me, I'd not even seen a dead one on the road. Which was what had reduced me to looking for chopped-up taipans among cane wagons.

Still, I could hope for better things inland. Inland was where many of the more recent taipan bites had occurred. These days, bites were rare in the littoral cane fields, not least because horse-drawn ploughs and hands-on machetes had long since given way to tractors and cutting machines. The automation of the industry in the late 1950s meant there had not been a taipan death in the cane fields for decades. Despite the enduring myth, the taipan threat had in effect migrated from the cane and was now present in a more generalized range of locations—and habitations. A little house motif appeared alongside the taipan information on my snake decoder; as the legend explained, the taipan was known to "come close to or into human dwellings." People were bitten in sheds and back rooms, in gardens and school grounds. They were bitten weeding round their properties or taking a swim. Nineteen ninety-six had already seen one taipan death when a man had leaned over to pick up a golf ball from scrub near Brisbane.

The road left the cane to climb through rainforest before emerging onto a sun-dappled plateau of pastureland and scrub. Eucalypt trees threw long shadows, bar-coding the road. Expanses of knee-high termite mounds and blackened tree stumps pinboarded the landscape. No More World Heritage, protested the signs nailed to the trees as the old copper town of Mount Molloy appeared on the brow of a hill. A prominent plaque commemorated pioneering prospector James "Venture" Mulligan. The manner of his passing seemed strangely inevitable. "Bought the Mount Molloy Hotel in 1903," the inscription read, "and died 24.8.1907 from injuries received when he tried to break up a fight in his hotel."

"Oh yeah?" said Mick. "Around here, people will tell you Mulligan started that fight." Mick was the publican at the National, Mount Molloy's homestead-style wooden hotel with wide verandahs and a corrugated iron roof that dominated the town much as the courthouse did in the towns of the American South. Mick was staring into a little pond, a feature of his alfresco dining area.

"Just put in a couple of barra," he explained. "They're the bigguns. They should eat the goldfish."

"Trying to restore the natural balance," I suggested.

"Natural what?" Mick wondered why the Pom was on about shampoo ads. "Trying to wipe out the little bastards," he corrected me.

Mick gave me a room on the first floor. It had old wood-

planked floors and walls, a ceiling fan and elderly items of forebears' furniture, dark chests of drawers that suggested Scarbor-
ough guest houses. On the wall, a single picture showed an
English huntsman spurring his mount towards a hedge, probably in Warwickshire. In the Queensland sun, his red hunting
coat had faded to a fetching pink. The furnishings confirmed
that Mount Molloy had had its time. Once, the place had had
the copper mine and a timber mill. It had even seen the
Wednesday goods train up from Cairns. It had had five hotels, a
butcher, a saddler, drapers and storekeepers, even an aerated
water manufacturer once.

My room opened onto the shaded verandah from where I
could see what little remained. The copper had given out, the
timber mill had burned down and the train had long since
stopped running. That left a handful of homes, the school, the
cricket pitch, the National, the phone box and a café up the
road called the Mount Molloy from where the fevered strains of
O Sole Mio could be heard. I pulled up a chair.

It was the lazy end of a hot afternoon. The sound of insects
was like a thick slurry, a sozzled orchestra sawing through the
heat. There was a slight breeze in the tamarind trees. Traffic
was slight; occasional, eighteen-wheel road trains thundered
past. Some, piled high with cane, were headed for the mill at
Mossman. Others were stacked with neat pyramids of timber
out of Cape York.

I could hear them minutes before they passed, a distant buzz
in the air that became a whine rising from the endless sea of
eucalypts and built to a roaring presence that shook the walls of
the hotel, setting my pink huntsman ashake and ruffling the
tops of beer glasses in the bar below. The piebald dog in the
road, I soon realized, was hearing the road trains even earlier.
The way his mothy ears pricked, rotating slightly like the radio
dishes of listening stations, forewarned you. As each juggernaut
filled the road, the dog rose casually to his feet, feigning un-
awareness until the vehicle had swept past. Then he suddenly

rounded on it, snapping ferociously at the rear axle, chasing the road train out of town in a cloud of dust from which he finally emerged, panting but satisfied at a job evidently well done.

A four-wheel drive heading north drew up at the National. A man disappeared into the bar. From my vantage point, I could see straight down onto his bald patch. At the patch's eye, where there was no hair whatsoever, it was rouged with an inflammation that he kept scratching. After he had been absent a few minutes, a woman leaned out of the passenger window.

"Chris," she whined at the pub. "Hurry up, will you? The kids are getting croaky."

From beneath the verandah, a clenched fist appeared.

"Well, the kids are going to get some of *this*," Chris replied.

The 4WD continued its way north, trailing traded shots— *fucking oven in here, Chris* and *You a girl or something? You could always get out, you know*—from the open windows. I went downstairs and made a call from the phone box. The strains of "The Stripper" playing loud attracted me to the Mount Molloy Café, where a few tables were arranged under brightly coloured awnings. The proprietor, who did not hear me enter, was playing an imaginary trumpet to the music. Then he began unbuttoning an imaginary top and slipped it off a provocative shoulder.

"Aha," he said, dispensing with the trumpet as he finally noticed me. He wore shorts and a singlet, had thick glasses and exuberant masses of curly grey hair.

"You wanna deluxe burger?" His accent was Italian. "Gorda," he yelled out back.

"Fat one Gordo yourself," said the woman, waggling a finger at his pasty stomach as she made her plump way to the stove, wiping her hands on her apron. She was dark, Mexican, happy and entirely correct about her husband. He was quite as fat as she.

"Hey," she asked. "You fed George yet, you waste of time?"

"George, George, George," Gordo replied, exasperated.

"George has eaten lettuce all bloody day." A sizzling rose from the stove.

"Hey, you like Elvis?" Gordo asked me and winked. He ran an imaginary comb through his hair, fitted a fresh C.D. and moved towards his wife.

"Hey!" she protested as he settled his hands upon her spreading waist. "I'm making the burger. Go dance with George, why don't you?"

Gordo disappeared to return with a young wallaby in his arms. George had been found beside his road-killed mother, and was clearly resigned to his rescuer's idiosyncrasies. As Gordo waltzed to the improbable strains of "Hound Dog," patient George nibbled at the fleshy ear that his dance partner presented to him as Gordo turned away for the big notes, an imaginary microphone to his mouth. Gorda flapped an affectionately dismissive hand at her husband and placed the burger in front of me. It came on a plate that she carried in both hands.

"Big burger," I said.

"Got to make it big," Gorda replied, wagging her finger. "Otherwise, how you spect people gonna come back." There was something splendid in Gorda's belief that her gargantuan burgers alone would keep people coming back, a heroic denial of the couple's geographical isolation, as if a burger run to the Mount Molloy Café was a reasonable option for most Australians. As splendid as the burger itself, so large that its contents were secured by a vertical skewer which I eased out to reveal layers of lettuce, a slice of beetroot, a fried egg, several slices of beef tomato, a pineapple round with a hole in the middle, onions, a puddle of gravy, an enormous patty of meat, and the same the other side in exactly the reverse order. I ate it as Gordo serenaded me with his favourite songs, "Una Paloma Blanca," "The Birdy Song," "Are You Lonesome Tonight?" and "Rock Around the Clock," then I wiped my mouth.

"Very good," I said. Gordo knew I was referring to the burger.

"Gotta be good," said Gordo. "Which reminds me!" And he turned to furrow through his C.D. collection. As I left, the strains of "Johnny B. Goode" rose into the sky.

The bar at the National was less flamboyant. There was a noisy clock, a slant of late sun falling across the pool table, a game of Aussie Rules on the telly and a couple of stringy old men in bush hats. Beneath the wide rims, where great clouds of tobacco smoke collected, their eyes were narrowed. Mick poured me a beer.

"And what might you be doing in town?" he asked.

I told him I was meeting a guy here.

"Oh yeah?"

"Calls himself Dundee," I explained. "Lives out at Mount Carbine. He got bitten by a taipan."

"You a doctor then?" Mick asked, looking alarmed.

"No, no, years ago," I reassured him. Dundee had survived.

Dundee bucked into the bar shortly afterwards, looking like the outback itself. His ripped bush shorts were stained with innumerable run-ins with what I took to be dirt, wood, leather and beer. A deep tide of fresh sweat had restored the original deep blue to the top of his faded singlet, and had also puddled in a series of circles, different-sized but remarkably vertical, down his front. Dundee wore a Stetson. His eyes were bright if beery, and his face was lined like old oak. I guessed he was in his forties.

"Pleased to meet yer," said Dundee, sitting down beside me. "And this is me cousin, Kev."

Kev was bald, but had a straggly red moustache and pirate's earrings. On his shoulder he carried, much like a bazooka, a four-foot didgeridoo to a fearfully casual effect.

It seemed that outback life had been designed for Dundee and Kev. Mick the landlord, by contrast, looked as if he had crash-landed here and was still adapting to the fact. Dundee had grown up in the Far North, and had spent much of his working life breaking wild horses through Arnhem land and the

Gulf country. The bar of the National Hotel was at the soft end of Dundee's spectrum.

Dundee got himself a beer. It was two years ago, he began. January, about the hottest time of the year. Dundee had been working over at his sister's place at Julatten, a tiny settlement just to the north of Mount Molloy. "It must have been late afternoon. I was spraying along one of Michelle's fencelines. She has trouble with lantana up there, see. That's a noxious weed that tends to do in the cattle if they eat it. I was in long grass, maybe knee high, and I was pumping to get the pressure up on this sprayer I was carrying when I felt this thing on my leg—bang! bang! two hits—and it just felt like I was hung up on a piece of barbed wire. So I pulled my leg away and looked down, and there was a fucking snake hanging off of it, see. Bit me through the jeans, on the left calf. So I jumped like nothing on earth and the snake fell off and headed into the grass. Now this was a dark-coloured snake, and I thought it was a red-bellied black. Now blacks aren't so bad, you know. They can give you a bad headache, make you sick. So I kept on spraying but then as I was working my way along the paddock, this thing moved out of the grass again, and I could see the full size of it. Over six foot, it was. And I said to myself, 'Hell, that ain't a black, it's got to be a king brown. Or a taipan.' "

Dundee's last word seemed to stir the old fellas. One of them looked up from his beer. The other swatted the smoke from his eyes for a better view. Mick winced. The unfolding story spoke uncomfortable volumes about the place he'd fetched up.

"So I stopped, and took off the spray container. Carrying it on my back I was, see, and it was then I first felt the bite burning. I remembered what they taught us about snakes when I was in the army. If you're bitten, they told us, don't panic. So I walked to the house calm as I could. About three hundred yards it was. By the time I got back my leg was tingling from top to bottom. But it wasn't, like, that painful."

One of the old fellas struggled to find a more comfortable

position. He had to lean forward to hear properly. His face, on which a lifetime's application had contrived an expression of supreme indifference, was threatening to break out into something approaching interest. I wondered whether the facial muscles could take the unfamiliar stresses they were being subjected to.

"The Rules," the other old fella said to Mick, gesturing at the TV. "Turn down the Aussie fucken Rules, will ya."

"So the first person I saw when I got back to the house was my 17-year-old nephew," Dundee continued. "I told him to get a bandage and bind my leg, from crutch to the bite just above the knee. Firm, I told him like we'd been taught, but not too tight."

"S' right," said one of the old fellas. "Firm but not..."

"So I lay myself on the couch, see, and then Michelle arrived. I told her I'd been bitten, and could she get on to the doctors, and you know she just laughed. She stopped laughing when I showed her the bite and the fang marks were an inch apart; it was a big snake had got me."

"Now that is a big 'un," said Mick, parking himself on a stool behind the bar.

"So it was while Michelle was talking to the ambulance," said Dundee, "that I started breaking out in cold sweats. All over, like. And that's when I first thought I was going to die, see, and I thought this is the way I'm going to go of all ways. A bloody snake kills me. Then Michelle's asking me the questions the ambulance man's asking her over the phone. 'He's asking how you feel,' she says. 'Are you losing your eyesight?' And all of sudden I am, you know. Michelle's a blur and out the window's just a blur. And all I know is the ambulance is on the way, and I'm losing it."

A man wandered into the bar. *Don't let your tool get blunt,* urged his T-shirt.

"Give me a XXXX Gold, mate," he said to Mick. But Mick stayed him with his arm. "Just hang on a sec," he admonished him. "I'm listening to a story here."

"Now, this is where it gets strange," said Dundee. "Suddenly, I didn't give a shit about dying. It was like a big high, as if somebody had hit me with a drug or something. Even when the ambulance bloke arrived, cut off the bandage with scissors and hit me with a big needle, I just felt like I was floating. It was a beautiful sensation."

"You got bit…" the man in the T-shirt began asking, but was interrupted by one of the old fellas.

"Taipan, fucken taipan," he told him before turning to Dundee.

"You're saying it was good, mate?" he asked.

"Good," said Dundee. "It was amazing. You know, if I was going to die of cancer or something and I couldn't take the pain, I'd get myself a taipan in a bag, and give him a little roughing till he bit me. That's the way I'd want to go."

"I heard that's what some blokes up in Mossman are doing," said Kev, making a few desultory sounds from somewhere in the tuba range down his didgeridoo. "You know, the real freakos. Getting some taipan juice, and injecting just a fraction. Gives them an incredible buzz, apparently."

"So then what happened?" asked one of the old fellas, getting impatient. "And you, cut the didgeridoo noises will yer. I'm trying to hear your mate." He turned to Dundee. "You having a beer, mate?"

"Thanks, I will. Well, all I know is what they told me later on. I was hallucinating all the way to Mareeba. I kept thinking there were all kinds of great aunts with me in the ambulance. I must have been in a bad way 'cos the driver radioed for a doctor to meet us on the way in. The doctor reached us at Biboohra, you know where that bacon factory is just outside Mareeba, and hit me with a big injection. Apparently, my heart had stopped so they got me going again with the electric shock thing, you know. So I guess we got to Mareeba at about five-thirty. Something like that. Gave me antivenom there. Then, that evening, they flew me to Cairns in a chopper. I'd never been in a chopper before but I may as well not have been on that one for all I remember of it. Apparently, I had diarrhea something awful at Cairns, and the doctors stayed up all night to keep me alive. I was in intensive care for three days. Then I spent a day in the wards. That's when Clive Brady came to see me. You know, the guy who got hit by a taipan in Mareeba a couple of years before me. Took seven bites, did Clive. We just sat there and compared notes, we did. Seems he had a worse time than I did. He even brought the video some American TV company made of the attack."

"And how are you doing now?" the man in the T-shirt asked.

"Well, they say it affects your sex life, snakebite does, but mine hasn't looked back since. In fact, being bit's done me no harm with the girls, I reckon. Hasn't done my nerves any good though. If anybody claps their hands or stuff, I jump. I wake up in the morning, feeling strange, insecure maybe…"

Kev sounded a long, sad note on his didgeridoo. His cousin wasn't one to admit to insecurity.

"Don't listen to him," said Dundee apologetically. He was addressing me, but a wider audience of bar staff and drinkers was now listening to his every word; the guy had survived a taipan bite, hadn't he? "We make didgeridoos together, Kev and I, when I'm not working at the wolfram mine out at Mount Carbine. When we're out bush looking for timber, I often wonder about snakes. If one of us was to get bitten by a taipan out there, there's no way we'd get to a hospital in time."

Kev giggled. "Tell them about our emergency pack, Dundee," he said.

"Nah."

"Tell them, Dundee. Tell them."

"Oh do," I said, interested that Dundee might know of some local herbal cure.

"You tell 'em, Kev," Dundee retorted, turning on his cousin. "Why don't you tell 'em?"

Kev looked crestfallen and wandered off, occasional didgeri-doo notes trailing after him.

"Oops," said Dundee after a moment, excusing himself. "One cuz in a strop."

Dundee didn't return. I called him up later that evening. Down the line, I could hear the sounds of the didgeridoo and what I took to be a chainsaw. The cousins were having a party.

"Where did you get to?" I shouted over the noise.

"Oh, you know," Dundee replied. I didn't. I asked him about the emergency pack.

"Oh that," Dundee shouted. "Well, if you must know. Me and Kev, see, we're out a lot cutting wood for didgeridoos. Way be-yond hospitals if a taipan gets us. So we take with us a joint, a XXXX stubbie and a *Playboy.*"

"And that's the emergency pack."

"Yup," said Dundee. "That and the shade of some beaut gum tree. We reckon we settle for a final nice joint, a final beer and a final pull on the old..."

"Thanks, Dundee," I interrupted him. I hoped it would never come to that, I told him.

I left early the next morning and nosed the car past the sleeping piebald dog. Even at this hour, there was music at the Mount Molloy Café. The low sun shone through a translucent awning to reveal two fat silhouettes, dancing the tango, as they awaited takers for their breakfast burger. George stood in the doorway, staring out at the road.

14. AFRICA IV

Vampires issue forth from their graves in the night, attack people sleeping quietly in their beds...those who are under the fatal malignity of their influence complain of suffocation and a total deficiency of spirits, after which they soon expire.

—John Heinrich Zopft, *Dissertatio de Vampiris Serviensibus* (1733)

JUNE 1996

Something woke me. Something in the room woke me. Which was no way to wake in Africa.

I lay motionless. Motionlessness seemed circumspect, at least until I knew what I was dealing with. I peered out through the mosquito net into the grey light of the room. Then I blinked, and looked again. At the foot of my bed I could plainly see a black mamba. It was enormous. It stood upright, its head six feet clear of the floor. At least, it was quite as high as the coat stand that usually stood there. A breeze from the low window shifted the mosquito net, lending the mamba beyond it a rippling, diaphanous aspect and accentuating the chainmail pattern of its scales, but I could see that the snake itself had not moved. It would not do so, I could only hope, just so long as I too stayed still. Move a muscle, and it was sure to strike with hammer-blow fangs, burying me in ripped lengths of mosquito netting and tinderized supporting frame.

Well, it wasn't an accident. It couldn't be. I'd heard African talk of snakes coming for people. Which was what this one had done. In the dark, like a night monster. Its half-hood hovering above me, it even looked like a vampire closing in on its prey, with its dark cape spread out behind it. It had stalked me, slid silently into my room at Plot 28 and was now choosing its moment to take me. I could barely breathe and my heart was racing, galloping along and kicking out at my ribcage. My skin prickled and dissolved into a film of chill sweat. As I braced myself for the attack, I felt my sphincter puckering as if it knew I might otherwise involuntarily evacuate my bowels. At least, my sphincter was looking out for me.

But the attack did not come, and I began pulling myself together. I began calmly assessing my situation. For one thing, the same wash of grey light from the window that had convinced me the mamba was no shadow also suggested an imminent dawn. I could smell roses at the window, and something else, maybe potatoes cooking in the staff quarters. Which meant help should not be long coming. For another thing, it was arrant nonsense to imagine this snake had been sent to get me. Snakes didn't do that; science said so. I must try to be European, rational. That the mamba had happened upon me was simply bad luck. It could have entered through a number of open windows. It was well known that mambas found their way into huts, outhouses, storerooms and even lavatories; the bad news was they proved themselves the worst of claustrophobes if they then found themselves sharing.

I was reminded of colonial officer John Crompton's encounter with a black mamba in a Zimbabwean hut. He had awoken to what he thought was the usual sound of the convicts sweeping the square.

"It suddenly occurred to me," he wrote, "that it was still dark"; and convicts did not normally sweep the square at night. "I lifted the mosquito netting," he continued, "and from the bed lit the candle on the side table. What I saw made me dive back into bed and tuck in the mosquito-netting. A large black mamba was slithering along all over the floor." It had got in, it seemed, by a small open window and was increasingly agitated by its inability to escape. Crompton lay in bed for three hours,

anxious not to attract the snake's attention. He was also worried that the mamba's movements might knock over the candle,
and so set light to the thatch hut. At dawn, his shouts eventually attracted an African sergeant. Forewarned, the sergeant
gathered his men, opened the door to the hut and stood back.
The escaping mamba died in a hail of blows from long sticks
and knobkerries.

I had an imminent dawn on my side, admittedly, but my
mamba was doing nothing so innocuous as flailing about on the
floor. It was standing over me, which was thoroughly unnerv-
ing. For its part, I wondered how long the snake could possibly
stay upright. I had been impressed, even as a child, by reports
of the mamba's ability to rear up which I had first read about
in the adventure yarns of Willard Price. Price's heroes, two
teenage boys called Hal and Roger who were forever off collecting wild animals, invariably record-breaking specimens, for
zoos, had captivated my boyhood.

In *Gorilla Adventure*, Hal and Roger came across what they
initially took for "a pole standing upright in the middle of the
road." It was only when the pole moved that the boys realized
they had disturbed a black mamba. It was, of course, a monster
of the species. It stood six feet tall while another nine feet of
snake lay in the dust. Which gave me some indication of how
much snake could be lying unseen at the foot of my bed, in and
around my luggage, among the guidebooks, the scorched notebooks, the dirty washing, the malaria medication. I could only
hope that this mamba was a little less ferocious than Hal and
Roger's which, infuriated by their vehicle's arrival, had struck
"the windscreen a terrific blow that cracked the glass and left it
dripping with poison." It then crawled through the car's open
window, "glided smoothly over Roger's back...and chose to
slide over Hal's neck." It was so typical of the way things always
worked out for the intrepid collectors that it then moved with
great compliance "into the sack which Hal had left invitingly
open to receive it." Bagging mambas was never so easy.

I lay listening to the sounds of the night, the broken squadrons of small flying things spiralling around the room. Like brain-washed pilots, they flew their hollow carapaces into the walls, and from the floor a furious buzzing ensued as they set about clearing their heads. None of which seemed to disturb my mamba. Stranger yet, however, was that it now seemed to have more than one hooded head. In fact, it clearly had several, perhaps as many as six, protruding from the snake's upper body in a regular arrangement, like petals. It resembled an ornate headdress such as an Egyptian pharoah or an Indian naga might have worn. This was clearly too ludicrous for words. I blinked with great deliberation, balled my eyes up tight to drive the sleep from them and dared myself to focus hard upon the snake. It was then that the many-headed mamba resolved itself into the missing coat stand, an old-fashioned, dark-stained one with steam-shaped tongues of wood for hooks.

Fuck it. I threw a sweat-soaked pillow at the coat stand, which swayed for a while. I exhaled as if for the first time in minutes, and hammered my fists into the mattress. The Lariam again. It was doing my head in. I had heard users complain how it gave them nightmares and depression. Others suffered from dizzy fits, migraine, nausea, panic attacks, epilepsy, loss of balance and even psychosis. I'd even heard of one young man on Lariam who unaccountably threw himself from the second-story window of a Nairobi hotel, and ended up a paraplegic.

My condition was less extreme. The initial symptoms bore a striking similarity to those of one user I had heard of. He had become obsessed by Lariam's anagrammatical possibilities, and in his hyper-associative mind was constantly rearranging the word's six letters in an eternally doomed attempt to make "malaria." But just as he was always one "a" short, so my persistent mental Scrabble-playing always left me with a spare "r"; for, true to my preoccupations, the word I continually found myself trying to make was "Lamia," the snake demoness of classic mythology, the prototype of the modern vampire.

In my case, the hallucinations came later. They were more akin to the experience of another user who, fearful of being robbed, had begun barricading himself into his hotel room after seeing people wandering the corridor in stocking masks. The Lariam seemed to be feeding upon my own fears in a similar manner, generating hallucinatory visions which I was clearly beginning to mistake for reality. I was seeing snakes everywhere. They'd begun by appearing on the edge of my vision. As I moved my head, I'd catch the tail disappearing from sight. Lately, however, they'd got more confident, moving from the periphery of my eyesight to insinuate themselves into what turned out, after extensive blinking, to be roadside sticks, a plait of hair, the head of a toothbrush, a discarded car exhaust and now a coat stand in my bedroom at Plot 28. It was as if I was living William Fitzgerald's Christmas Day of 1891, when he saw snakes everywhere he looked, only Christmas was every day in my case.

"Barbara?" I asked, clearing my throat as she finished listening to the World Service news, part of her morning routine which did not brook interruption. "Do you mind awfully if I move that coat stand from my room?"

She observed me curiously over the breakfast table. It was not often guests felt the need to rearrange the furniture at Plot 28.

"Doubtless you have your reasons," she replied, spearing a mango cube. At that moment Ferdinand poked his head round the door. He looked worried.

"The m'ganga fish man has come back," he told her ominously. "With three more lobsters."

"Oh, Ferdinand!" she exclaimed. "Somebody must have seen you releasing that lobster on the beach. Tell him that our white mchowis are very strong, and we'll put a curse on him if he goes on treating us like this. We will not be blackmailed for our love of animals, not even lobsters. We cannot afford to buy all the lobsters he can catch just to return them to the reef for him to catch them again. Tell him to take them away. And if he wants my business, he'd be wise to take them back to the water. That's sea water, not cooking pot water." Barbara shook her head with frustration.

"And what are you doing today?" she asked me.

I told her I had an appointment with a mchowi, but promised to be back for dinner.

"Well, ask your mchowi if there's anything dramatic he can do with lobsters, will you," she replied, turning to the window to watch the shamed m'ganga slipping into the trees, Bumble trailing at his toes.

Jacob was waiting for me at the Kilifi bus stop. He was neatly dressed, even by his fastidious standards. His freshly pressed sleeveless blue shirt indicated that it wasn't every day Jacob met mchowis.

"I have two for you," explained Jacob in clandestine tones. "They are partners. We made contact through a friend of a friend." Mchowis didn't advertise their services, didn't leave calling cards. But there were always people who knew how to find them.

"My friend said they were suspicious at first. But they finally agreed to talk. For a consideration, you understand."

Jacob led the way. Kilifi lay under a canopy of acacias and baobabs, pooled in shade. It was a sombre town; the nearby ocean was nowhere. Even the fish, piled on stalls and swathed in spinning flies, seemed to have travelled a long dusty road to get here. Ubiquitous postered notices advertised an imminent prayer meeting with a Reverend Rejoice. Women sat on the hard-baked earth outside their shacks, making baskets from raffia. Men emerged from blackened workshops, their ragged clothes covered in oil, holding unidentifiable car parts. Goats cruised the roads, picking at discarded rinds of coconut. Casting furtive glances behind him, Jacob led me past a mosque and down a red-dirt path that skirted fields of maize and stands of mango trees on the edge of town.

Soon, the sounds of the town fell silent. In the leaden heat, there was only the drone of insects, birdsong and the trudge of our own feet. Ahead of us, a single wooden hut appeared among fields on an island of hard, bare earth. Some grimy children were slumped on an old bed frame under a cashew tree. One of them, who had been sleeping, woke at our approach; the frame mesh had left a diamond pattern imprinted deep upon her face. One such crevice funnelled a single tear diagonally across her cheek. At Jacob's prompting, all the children but she ran off in search of their parents. Thick with sleep, the girl steadied herself and watched us, occasionally brushing the flies from the snot at her nose.

Soon, a man and a woman appeared, so silently that I did not notice them until they were settling among us. They had arrived from different directions but reached us simultaneously, trailing children in their wakes. They did not speak to each other, but occasional glances passed between them that suggested a highly developed, even a psychic understanding. Otherwise, there was a conspicuous absence of the mumbo-jumbo I suppose I had expected; no potion bowls, strange vials or weird fetishes lying about, nor any incantations of apprentice initiates sounding from among the trees. They didn't appear to the sound of drums, in masks and headdresses, with lion pelts on their backs, filed teeth, dramatic earrings and armlets; these were not giveaway witch doctors. Plain-clothes mchowis, they wore the same discarded Western clothes that you saw all along the coast. He was in tattered shorts and a singlet; she, in a long skirt disintegrating along the hem and a grimy blue and white striped T-shirt. Beneath her eyes the decorative arcs of the Giriama had been scarred.

The real clues were in their hair—so explosively unkempt and matted, even by local standards, as to suggest minds differently ordered—and in their eyes, at once visionary and vacant. Sometimes their eyes seemed full of loopy depth; at other

times, they were just ballbearings in sockets, milky and sight-
less. For a long time the couple looked at me. I assumed they
were reading my inner secrets and bore their close scrutiny
with equanimity. Finally, when the man spoke, he seemed
strangely impatient.

"They want to know what they can do for you," Jacob translated. "They have work in the fields to return to."

I had clearly misunderstood. I suppose I was expecting the
usual elaborate sorcerer's hocus-pocus, but was instead being
minded to understand that the couple had business elsewhere
to attend to. And that it would be appreciated if I got on with
whatever I had come for. The implication was that I had been
granted an interview, not a magic show. These people felt no
need to impress me with a demonstration of their tricks. They
were simply here, as requested, to discuss their work. For that
was how they regarded the practice of their powers, as a commonplace everyday activity. It was strange enough to infer from
this that they had absolute belief in their powers. It was profoundly shocking to also realize they had no sense that I might
possibly doubt those powers. Around these parts, it seemed, nobody ever did. In Kilifi, magic was unexceptional. Magic happened. And so I started asking the couple about their jobs.

Through Jacob, I tried to reassure them I intended no harm.
Nor did I wish to steal their knowledge. I wasn't interested in
seeing them at work. I only wanted to know what they did. For
a living. The couple looked at me. The ensuing silence threatened to derail the whole process.

"Down payment," the resourceful Jacob whispered, remembering his diploma in business admin. A few hundred shillings
seemed to do the trick. The couple explained that they both had
the power. Like the Watamu m'ganga, it had begun when they
had seen snakes fighting. Like the m'ganga, they had learned to
cure snakebite. But that was easy. They could not say what it
was that had led them into the darker, more difficult areas to

become mchowis. There had been strange dreams. Sudden promptings whose origins they could not credit that had drawn them into more shadowy accomplishments.

"So," I asked. "What would those accomplishments be?"

"People approach us," said Jacob, translating for the woman. "We'll hear from somebody we know that somebody would like to visit us. To discuss a grievance. The meeting is arranged."

I could not understand the woman, who was speaking Mijikenda, the local language, but her casual manner conveyed the notion that she was merely stating her terms of business.

"Maybe somebody has been offended over a land issue," Jacob explained. "A boundary dispute. Or there has been an insult, a bad insult. Perhaps two brothers are having a relationship with the same woman, and her family is gravely offended. But they have to have good reason for coming to us. Even if somebody offered us lots of money to harm somebody, we would not do it unless punishment was deserved; besides, our dawa would not otherwise work. So we will listen to the grievances of such people. Then we will talk with the local elders. If they agree that there is justification, then we will go to work."

"Go to work?"

"Yes."

"What does that mean?" I insisted.

"We send snakes." The woman spoke evenly. Her delivery was matter-of-fact, but her words caused the older children to look up sharply; so this was what their parents got up to.

"Sometimes," she went on, "we send magic snakes. We provide the wronged person with a piece of bark. We tell them to lay it in the path of the wrongdoer. We teach them a chant they must sing over the bark to give it the power to bite. After a few days, once we expect the bark has bitten the wrongdoer, we ask that they collect it and return it to us. We place the bark under a special tree and wait for the tree to wither. When it does so, then we know that the spell is working. It tells us the victim is sickening. We must continue to watch carefully, however. If the tree then begins to recover, we know that a m'ganga is working against our dawa, and stronger witchcraft is necessary."

I wondered what stronger witchcraft entailed.

"It's exhausting," said the man, tapping his head. "Very hard work." He might have been bemoaning the daily commute, or the seasonal rush in the office, or a backlog of presentations that was taking him all over the country, hammering the mileage on the company car. Stronger witchcraft actually entailed going into the forest to whistle. They whistled for snakes—real snakes this time. Like this, he said, and whistled twice. The sound was short, like the report of a gun and quite without melody.

Once the snakes appeared, the man explained, one was selected and dispatched amidst chants with instructions to seek out and bite the wrongdoer. Mostly, they used mambas, the man said. Mambas were the worst of the snakes. But since the best m'gangas were quite used to curing mamba bites, they occasionally sent other snakes, notably cobras, to confuse them. I asked the couple how they knew their magic had been successful.

"We hear about the death from the hospital," she replied blankly, as if so much were obvious. "Or in the newspaper."

Snakes, it seemed, really were sent to get you in Kilifi. They were local instruments of justice and correction, delivering the death sentences that public opinion handed down. So my skin prickled and I even wondered whether it really was a coat stand I had seen in the night. And for a moment I thought I had to ask the ridiculous question that played across my lips—*you didn't by chance send one last night, did you? To Watamu?*—before I came to my senses, and it died in the asking.

"Final settlement," Jacob reminded me as we rose to leave, and I peeled off the agreed sum. The man pocketed the notes, then raised a finger to his pursed lips. The woman did the same and the children followed suit. The children then accompanied us to the edge of the clearing, where the path ducked into the fields. I turned to wave at them, but the children were already looking back at their parents with new respect. By the path, a tethered goat was giving birth. It stood patiently, its hind legs quivering. The kid's wet, matted head was visible in a tangle of blood-strung mucus. A rainstorm threatened.

I wondered what all the secrecy had been about.

"They're scared," said Jacob. "Whenever anything goes wrong in Kilifi, people blame the mchowis. A bad day at work, a boil between the shoulders, weevils in the maize, the mchowi is responsible for all these things."

Local laws had been passed to suppress the mchowis. In 1977, the district commissioner of Kilifi had banned traditional dances, considering them apt to "instill fear and devilish activities such as witchcraft." Often, locals took the law into their own hands. A neighbourhood mchowi had recently been severely beaten by a gang of youths in the aftermath of an unaccountably sudden death. Just last month, another had been set upon by his own nephews, and had had all four limbs broken. Others had been hacked to death. Killing mchowis was nothing new. In Zimbabwe and Angola, they had tended not only to kill them but to burn their bodies and scatter the ashes to prevent their spirits returning as avenging snakes.

"Of course," said Jacob, "you can't blame the mchowis for everything."

I chuckled, assuming this educated young Kenyan was being ironic. A good sceptic. For Jacob spoke foreign languages, and had studied business administration. It had not yet dawned on me what Jacob actually meant; that while mchowis were behind much of the trouble in Kilifi, you could not blame them for every last scrap of it. Jacob believed absolutely in the power of mchowis. He merely conceded some misfortunes might not be of their making.

"Of course they send snakes," said Jacob. "Everybody knows

they do. The problem they've been having with their identity cards! These days, everybody has to carry such cards. Every time the mchowis have their photos taken, the pictures keep coming out blank. The photography studio here just doesn't know what to do about it."

FEBRUARY 1997

"Superstitious!" said Pete, shaking the last few drops from the coffee pot. "You can say that again. The Kenyans see shaitani, ghosts in every shadow. Back in 1929, a few months after I was born, an earthquake hit the farm in the Solai Valley where my parents were employed in the middle of the night. A hole several feet deep appeared in the family cabin. In the morning they sent the African boy up to the owner, Colonel Lane's place, to get some milk. He returned soon afterwards with an empty pail, a milky pallor and much talk of shaitani, saying the colonel's place had disappeared in the night. We never saw him again." It transpired that the earthquake had thrown up a hillock in front of the colonel's place, doing no damage to the immediate house but ruining the sundowner-soaked sunsets from the verandah thereafter and putting an awesome fear into one young African.

"It was the snakes, of course, that scared them more than anything," Pete continued. "There's a snake they call the *ngorloco*, the Western sharp-nosed snake to you and me. Quite common on the coast. It's venomous but only mildly so. Very unlikely to do more than make you sick. Still, I've known families simply move out for good if one of them got into the house. Take their belongings and flee. Then there was another chap in Kilifi, a bit of a drunk who took a puff adder bite on the ankle. Got gangrene around the bite and was forever being fixed up at the hospital. They amputated at the ankle but the gangrene kept spreading. So they amputated at the calf. Still, the gangrene spread so the fellow was forever having his wooden leg extended. The Africans were terrified. I think they thought the snakebite was turning the poor chap to wood."

In the quiet of the late afternoon, the hotel bar had emptied. The jukebox played a slow number. A maid was hoovering in the hall. A waitress was clearing the tables, emptying ashtrays and collecting plates where mustard had dried into smeared hillocks. Our talk had become tentative as we felt our way towards the story I had come for.

"And the last time," I spoke eventually. "The last time you were bitten?"

"Excuse gents," an apologetic voice interrupted us. It was the maid. "You won't mind if I hoover around you. We've got a reception here this evening. Just lift your feet if you will, dear."

So Pete Bramwell tucked his legs up into his chest and, over the sound of a hoover, began telling me about the time he was bitten by the black mamba.

"It was one day in April," he began. "A hot morning with great thunder clouds bubbling up. I was at home at Takaungu doing chores. It began with Jan calling out from down at the rabbits. We kept them in a run which was shaded by a large open-sided makuti thatch structure. And mambas just loved that thatch. Well, that morning Jan saw one there. So I went down for a look and then went off for my catching tongs and a bag which I'd left in the back of the truck. Well, my first—"

"If I can just get round the back of you, dear," said the woman with the hoover, "then I'm all done."

"—my first mistake was to miscalculate the length of the snake. Much of it was hidden in the thatch, and the bag was too small. I also got a bad hold with the tongs. I should have secured him no more than a foot from the head; in fact, I'd left as much as three feet free. At that point, I should have let him go, stood back and started over. But I kept on with it. Still, I got him in the bag, somehow holding the tongs and the neck of the bag with one hand and the drawstring with the other. I released him but I wasn't quick enough. The mamba was out before I could get the drawstring shut. It got me on the forefinger joint, twice. I didn't panic. I didn't need an angry mamba around so I killed it. Wasn't vengeance. Just good sense. Smashed it against some rocks. Then I walked up to the house. I took it slowly. Jan had gone up to the house; she had never liked watching me catch snakes."

" 'It got me,' I told her but I think she knew even as I walked in. I must have looked pretty pale. Jan sat me down on the sofa and drove off to find our good friend David Corroyer. David had a very fast car. Japanese, I think.

"I had some antivenom at Tak. Forty ccs, I think, which I injected immediately. But I knew that forty ccs wasn't going to be enough. The snake had got me good. Bites didn't come worse than a black mamba to the hand. I knew I was in deep trouble. I sat on the sofa, waiting for Jan to come back. Pretty soon afterwards, the nausea hit. I was retching bile into a bowl. I began sweating and breathing quickly got difficult.

"I tried focusing on a wall calendar to collect myself. Which was when it struck me. April 24, 1977; it was a Sunday. Jan had always had a thing about Sundays. And years ago, I had made a solemn promise to her that I would never catch snakes on a Sunday. I hadn't meant to, but I had just broken that promise. And look what happened."

Bagging mambas never was that easy.

JUNE 1996

I left Jacob and Kilifi after lunch in a rainstorm. The bus threw great, serpentine slaloms to avoid the puddles on the coast highway. I picked up the Double Happiness matatu on the final stretch down to Watamu. The driver had exciting news.

"Caught a big, big fish in the night, and so they did," he told me, extending one arm through his window and the other past the noses of the several front-seat passengers. So I left the Double Happiness at Hemingway's for a look.

"A great white," said the proud receptionist, pushing open the hotel's cold-room door. "Only the fifth one caught off Watamu in living memory." A huge shark's head about four foot square lay upon the slab.

"There's no way they could have landed the whole thing," the receptionist explained excitedly. "The head alone weighs a thousand pounds. Must have been a five thousand pounder all in. That's about as big as they come."

The shark's head lay on its side. Its flesh was grey and rubbery. The baleful eye that stared out from its prow profile looked no more dead than it would have done in life, which, I mused, was the essential strangeness of sharks. Its mouth lay slightly open, exposing layers of triangular teeth. Serrated and translucent, they overlapped along the jawbone like a childish depiction of the Alps. Lightly, I ran a finger along the peaks.

"You here for the fishing?" asked the receptionist as we stepped out into the heat.

"Strangely enough," I said, "I'm looking for black mambas."

"You are?" he grimaced. "Horrible things. I'd rather take my chances with a great white any day. There was a bloke down this way, you probably heard of him, a snake collector, must have been twenty years ago now. Got bitten by a black mamba. He was in hospital for the Lord knows how many weeks. Was paralyzed but, you know, he could hear everything. Even heard the nurses discussing the size of his, er, you-know-what! None too impressed, apparently. But that wasn't the worst of it. He also heard the doctors telling his wife that she should expect him to be extensively brain damaged. Even suggested turning off the life-support might be the kindest thing. Can you imagine? There was nothing the bloke could do."

"And you wouldn't remember his name," I said wearily, knowing the answer.

"I wish I could remember," the receptionist replied. "Barrett. Was it Barrett?"

The taxi-driver who took me back to Plot 28 in the darkness suggested dropping me at the top of the track; he didn't fancy the thick sand. I insisted on being taken to the door. Over dinner, Barbara was sorry to hear they'd killed the great white. It was a stifling night, but I closed the windows when I went to bed.

15. INDIA II

The occasional expansion of the skin of the neck in the form of a hood, ascertains the cobra's identity to the most superficial observer.

—Patrick Russell, *An Account of Indian Serpents Collected on the Coast of Coromandel* (1796)

AUGUST 1996

The man had made a stage of the sandy scrub behind the Lakshmi Lodge. He sat cross-legged, busying himself with his kit; the round, flat reed basket, a drawstring bag, a roughly fashioned flute pipe and the man, an auxiliary of sorts, who sat by him. At their backs, the line of surf curled, collapsed and gargled a thin tumble of pebbles. The beam of the lighthouse perched on high ground south of the town sliced into the deepening dusk. It swung along the beach, illuminating for an instant the stepped towers of the shore temple, carious with erosion now, the mobs of strutting seagulls and the dusty-dry casuarina pines. It lit up the fishermen's beached boats, open rafts lashed together from baulks of Kerala pine, their hulls like celery stalks, bowled and widening towards the stern, and a rising prow that had been added to the pointed bow. Occasional waves reached the rafts, shifting them lazily on the sand.

I trudged through the scrub towards the two men, who feigned not to notice my approach until I was standing by them.

"You came," the main man shouted, as if to make himself heard above the crowds he dreamed of drawing; as if I had only just made it for a performance that would wait on nobody. The

man straightened his kit again, meshed his fingers in a mae-
stro's preparatory flex and looked around him. Where I saw a
long beach and the pale figures of fishermen scavenging drift-
wood along the shoreline for evening fires, he seemed to see the
desperate scrabble of admirers' hands for tickets bearing the
stage name they had bestowed upon him: "Cobra Man" or
"Snake Man" or "King of the Reptiles." Mamallapuram's, no,
Tamil Nadu's—nay, all India's first and finest snake charmer.

Nothing could keep him, however, from finally acknowledg-
ing that I was the sum of his audience. He introduced himself
with an ill-advised theatrical sweep of the arm that caused him
to stumble.

"Mr. Emam," he pronounced. "Snake charmer."

Mr. Emam, snake charmer, smelt of drink. He snuffled,
smoked incessant cigarettes scented like smouldering rubbish
tips and had bloodshot eyes.

"Our father was snake charmer," he proclaimed. "Our grandfather too. Charmed snakes for King George. And was a snake
charmer all over Germany. Leipzig, Hamburg, Dresden..." Mr.
Emam then hiccupped. A glorious lineage had evidently done
nothing to deflect a parlous present, which was hiccups, cheap
drink and the unremarkable companion to whom Mr. Emam finally gestured.

"Mr. Emam's brother," he said with scant enthusiasm.

Mr. Emam's brother had a discoloured right arm, patterned
like a Dalmatian dog's coat. Strangely circular islands of brown
pigment survived in a rising sea of sickly, ivory-coloured skin. In
the brother's hairstyle, you could just about recognize evidence
of the ordered, wavy corrugations that brilliantine had brought
to it at some stage in the past, some weeks before, when a possible job had beckoned or a one-time friend with rare influence
had briefly returned into Brother Emam's fallen orbit.

For the Emam brothers, things had come to a sorry pass, and
I determined at least to prove a sympathetic audience that
evening. I followed the brothers' example and sat on the sand

opposite them, legs crossed. Together, we formed a rough triangle around the reed basket. For the moment, however, the
basket remained untouched, a promise of things to come, as
Mr. Emam preluded with his bag of tricks. He pushed back his
sleeves and removed from the bag several brass pots which he
arranged before me. Magic scarlet balls moved between them
with a sleight of hand certainty that belied Mr. Emam's drinking. Beach pebbles spewed from his mouth. He set fire to a
length of cotton thread, then hauled it, fully restored, from his
left ear.

To indicate that he was ready for the main attraction, Mr.
Emam extended his arms with the gravitas of a classical con-
ductor. The brother lifted the flute pipe in both hands and
passed it reverently to Mr. Emam. It was made from a dried calabash gourd about eight inches long, with a blow hole at the
thin end. From the gourd's bulbous end extended a bamboo
length in which finger holes had been burned. The pipe had
been oiled with linseed, and was decorated with a rough circular inset of red beans.

Mr. Emam began to play. At the first sound, the brother lifted
the lid from the basket with a waiter's flourish. And the cobra
within it rose instantly, its heart-shaped hood swelling like a sail
catching the wind. All up its narrow belly, arched horizontal
lines showed across its ivory skin that somehow reminded me of
the segmented trunk of a bamboo tree, swaying in the breeze.
Steady, I told myself.

The pipe notes were crude—they had a bagpipe stridency—
and the player's drunkenness was showing now. But the tune
still achieved a kind of stature. It suggested the big oriental
themes, the final showdowns and divine displeasures that are
staples of the great Hindu epics. It was a fitting soundtrack to a
classic Indian cameo, one repeated time and again until it had
acquired a shorthand symbolism for the continent's unknow-
able heart. For as Mr. Emam swung the pipe from side to side,

so the snake seemed to move in time to it, a slave to the rhythm. Here, in this brief communion, was encapsulated all the mystery and wonder of India revealing a mystical relationship between the cobra and the music of its charmer, a connection reached outside language or intellect, an audience with the gods.

Or so it was claimed. Sceptics, of course, would have it otherwise. In India, where much seems inexplicable, accusations of sleight are never far away. From India to Egypt and Morocco, the snake charmer has long been synonymous with the showman and the sorcerer, the mountebank and the cut-price illusionist, using every available trick to suggest preternatural snake skills at no personal risk. Typical was a noted charmer of Luxor, Egypt, who used to attach himself to tour parties, claiming to be able to smell nearby cobras along the way. All at once, he would inform the tourists that he had scented a cobra and, as if by magic, would proceed to haul one from a hole in a nearby wall. The cobras invariably submitted to the ordeal with appropriate irascibility; evidently, they had been enduring it on a regular basis. What the cobras had not yet learned to do was bite through the cord with which the snake charmer had surreptitiously secured them to the wall during the early hours.

Few charmers, however, went to such elaborate lengths. Most simply rendered their snakes innocuous. The theories were varied as to how they achieved this. In the late 1700s, the Abbé Dubois, a renowned observer of Indian life, claimed that snake charmers "take the precaution to excite the snake every morning, forcing it to bite several times through a thick piece of stuff so that it may rid itself of the venom that reforms daily in its fangs." The snake-fearing Gordon Sinclair believed that charmers protected themselves from accident by tranquillizing their snakes. "Here's a scrawny lad with his basket of cobras as a pillow," he described a Bombay scene. "When dawn comes he'll bathe at a water fountain, dope his snakes a bit to make

sure they're harmless and prowl the streets looking for breakfast." Other charmers, it was claimed, took the more direct
route of simply sewing up the mouths of their snakes.

Mr. Emam had been keen to preserve the secret of his magic
tricks, challenging me to explain his moving red balls, his mul-
tiplying beach pebbles and his miraculously restored cotton
thread with a series of interrogative *huhs?*, but he was strangely
revealing about his cobras, as if they had their own, inherent
magic.

"We just break the fangs off," he said. "Snip, snap. Of course,
they grow back again. Lest we forget—my finger nails! Every
time the nails grow this long, then is the time to break the cobra's fangs again. Once, we used to leave their fangs alone. The
audience liked that. But there was no danger from the cobras;
we had taught them not to strike."

I wondered how they had done that.

"Saucepans! We made them strike at hot saucepans. Soon
enough they got a jolly hot nose. After that, they didn't strike
any more. Well, not often anyway."

Legion, then, were the tricks of the snake charmer's trade,
but it was in the act itself, in its very name, that the greatest de-
ception lay. The cobra could not be charmed by Mr. Emam's
music since, as any scientifically minded child will tell you, all
snakes are deaf. Since snakes cannot hear the tunes of their
charmer, no relationship exists between the music and the
snake. The claims of snake charming make as much sense as
movies for blind people, FM stations for the deaf.

I remembered how African mambas reared up and threw a
hood of sorts. This they did when they were angry. What was
passing between Mr. Emam and his cobra was not mystical communion, then, but confrontation. The snake was not
dancing to Mr. Emam's tune; it was giving him due warning.
Charmed? The cobra was infuriated. Even as Mr. Emam's
brother had lifted the lid off the basket, I remembered how the
cobra had risen with unmistakable intention, less like a dance

partner from a table than a boxer from a corner, causing my saliva to slide chilled past my Adam's apple. The only relationship this snake was interested in pursuing, it seemed to me, was of the eyeballing outside a Glasgow pub late on a Saturday night kind. It was indeed responding to Mr. Emam's every lead, but its mirror moves were simply defensive adjustments that best readied it to strike.

By now, however, the performance was beginning to unravel. The drink had inroaded Mr. Emam's reserves of concentration, causing rogue pipe notes to flit into the night. The cobra was also beginning to tire. Most likely, it had not eaten since its capture weeks before. As Mr. Emam lay down his pipe, the cobra sank gratefully back into the basket. Which was when a strange thing happened.

In an instant, the cobra had lost its height and its hood, and its spectacle marking shrivelled from sight, like the promotional message printed upon a popped balloon. In an instant, the cobra had transformed itself into an indistinguishable brown snake. It was no longer recognizable as a cobra. Indeed, there was a powerful sense in which it no longer *was* a cobra, as if the classic cobra form, the hooded rearing stance born of anger or excitement that has rendered it an icon, an icon of absolute intent, was the only true version of this snake.

Mr. Emam's brother dropped the basket lid upon the deflated snake, and payment time loomed. Passing over a few rupees, I asked Mr. Emam why he had stopped working with fully fanged cobras.

"I think the cobra forgot the lesson of the hot saucepan. My brother was bitten," he replied, thumbing at the brother's discoloured arm. "And getting bit is not worth the trouble," he continued. "Not," he added, pointedly examining the bank notes, "for the money we get, mister."

In the morning, Ramesh, the young man at the Lakshmi, approached me across the roof terrace, a breakfast tray laden with toast, tea and mango resting on one upturned palm. The fruit had been cross-hatched, then turned inside-out to form a neat, mango-plated hemisphere. With his free hand, Ramesh executed a sinewy, serpentine cobra performance; he had evidently been keeping tabs on my evening.

"And how is our *Pamu* Emam?" he asked me, using the Tamil word for snake charmer.

He seemed ill, I told him.

"Fever, fever," said the sympathetic Ramesh.

"Booze, booze," I replied.

Ramesh looked horrified. "But Pamu Emam's grandfather was snake charmer to King George," he said, as if such an appointment must put him above intoxication. "Then his brother was bitten. Things are not so good since then."

"Tell me, Ramesh," I said. "What do Mamallapuram people do when they get bitten?"

"Some do nothing," said Ramesh. "Some of them die. And some go to special snake surgery, Tirukalikundram."

I hired a moped and set off for Tirukalikundram that same morning. The road passed through wheat fields under a sky patterned with swirling wisps of high cloud. Ten miles inland, the small town of Tirukalikundram was famed for its hilltop temple where holy vultures en route from Varanasi to Rameshwaram in the far south were said to alight at noon every day, an event which attracted an expectant crowd of pilgrims, tourists and onlookers. I was more interested, of course, in the town's special snake surgery, which was said to attract bite victims from miles around. Its reputation made the local roads notorious for speeding snakebite transports—requisitioned trucks, mopeds and even tractors—especially at night, when most bites occurred.

"During my whole residence in India," wrote the Abbé Dubois, a Mysore missionary in the eighteenth century, "hardly a month has passed without some person in my neighbourhood suffering sudden death by the bite of a serpent." Even today, it is estimated that 10,000 people die annually of snakebite in India. India has an immense rural population. It also has a lot of venomous snakes. That one of the most venomous species doubles as an object of worship, honoured wholesale at holy times and largely tolerated whenever it chooses to enter the home, rather compounds the problem. Nineteenth-century reports of householders doing *puja*, or offering worship, to the same cobra that had just bitten a member of their family exemplify remarkable attitudes that prevail in the late twentieth century.

Such reports were originally solicited by Sir Joseph Fayrer, a renowned surgeon at Calcutta Medical College. To establish the extent of the snakebite problem, Fayrer had sought relevant information from a range of civil surgeons, medical officers, assistant apothecaries and commissioning officers all over Bengal. He described the conclusions of his investigations as "perfectly appalling," for Bengal in 1869 alone, he counted over 6,000 snakebite deaths. The victims included grass cutters and labourers, water carriers and customs officers. People were bitten collecting firewood or returning home through the fields, answering calls of nature outside their Calcutta shacks or even lounging in airy billiard rooms.

Occasional, occidentally minded doctors reported that they had conducted post-mortems where embolisms, kidney congestion and pulpy spleens indicated snakebite. More often, however, cause of death was ascertained by traditional methods. Single hairs might, for example, be plucked from corpses' heads; those that offered no resistance were believed to confirm snakebite (rather than incipient baldness) as the cause of death. Where snakebite was suspected, corpses were often dispatched into the river in the Hindu tradition, and cause of death was unverifiable.

Many snakebite treatments were similarly traditional. The cure-all administering of "chants and incantations" was popular. A Brahmin priest might be summoned to breathe over the patient. Water was sprinkled upon the eyes seven times, prayers recited. Patients were revived from their envenomated torpor to smell remedial leaves. The leaves of *nim* trees, a renowned snake repellent, were crushed with chillies and administered internally. Alternatively, the victim's body was swept with a broom of nim branches to disperse the poisons. Cakes made from charcoal and cow dung were rubbed upon the wound. According to the Abbé Dubois, it was the custom that the victim of snakebite was lain upon a bundle of reeds and upon him was placed "a leaf on which is written a blessing for that person who will accidentally light upon him, and save him by a charm from destruction," which sounded less like treatment than an admission of defeat combined with a long-odds appeal for help. It amounted to an inexact science; even the more progressive doctors often relied on large amounts of brandy as treatment.

Locals had customized the road into Tirukalikundram as a threshing floor. To avoid leaving tire marks over the corn harvest or over the threshers, I followed a considerate truck driver into the trees where our minor convoy immediately got lost. The truck driver looked back to reassure me with a dazzling smile before climbing onto the roof of his cab. From his vantage point, he could evidently see how best to regain the road.

"Aha!" he pronounced and descended with renewed enthusiasm to lead me a charge through a backyard where he stopped to work a washing line clear of his cab before emerging with a triumphant hoot of his horn on to the road again.

In the middle of town, by the Jeeva snack bar, a throng of snake stones surrounded a statue of Ganesh, the portly elephant of prosperity, all in the broad shade of a pipal tree. At the foot of each stone a pile of rice had been left. The snakes, smeared in turmeric, had been carved in a range of representations. In one, a pair of snakes appeared in profile. They were arranged vertically, their intertwined coils forming a series of circles which enclosed lotus flowers and images of Krishna. At the top of the stone, the cobras met, their hooded heads eyeballing each other. Other stones portrayed the snakes head-on, their great cresting hoods dotted with scarlet *kum-kum* powder.

The snakebite surgery was tucked behind the enormous brown temple that dominated the town, on Big Street. No chants or incantations sounded from within. There was, however, plenty of tubercular coughing. "Dr. B. Thaniga," read the painted sign outside. The sign also featured a swaying cobra, another snake I took to be a common krait, a scorpion and an Alsatian whose tongue was hanging out in a rather more approachable manner than the artist may have intended. Inside the surgery, a dark room with benches along the walls where mothers, children and sorry-looking men crowded, photographs of the same creatures hung from the walls. There were many more such dogs; again, these were groomed Home Counties beagles and spaniels, more likely to lick you to death than to subject you to a rabid savaging. There were also photos of kraits and cobras, Russell's vipers and saw-scaled vipers. Bites, it was clear, were the surgery speciality; coughs, it seemed, plugged any remaining gaps in the appointments book.

Even the creatures, however, gave precedence to a gilt-framed portrait of Shiva, who shared pride of place with a sepia photo of a dapper man in military uniform.

"Barathi Thaniga. My father," a voice remarked in English. She sat behind one of two desks at the far end of the room, inspecting a distressed girl. "He was a colonial officer in Pondicherry. And an expert in the traditional treatment of snakebite. He died some years back. Now my brother, my husband and I run the practice. My name is Shanthi."

She paused to peer into the girl's eyes.

"Did something bite you?" she asked me.

I was just curious, I told her. I wondered how many cases of snakebite the surgery saw.

"Oh, oh, oh," she muttered alarmingly. "Between March and June, the hottest and driest months, at least one a day. Sometimes several. Few of the people wear shoes, you see. Many are bitten in the fields, and many at night. As you'll see from our

admissions register." She handed me a red book which was littered with two A.M. entries. During July, which had just ended,
I counted up the cases of snakebite the surgery had treated:
seven of cobra, four of Russell's viper and five of krait.

"Many of the patients are here for fever, mostly malaria," said
Dr. Shanthi, testing the girl's blood pressure. "But that chap,
the quiet one, he's in for snakebite. Russell's viper. Happened at
dawn. A virulent, strange poison that lingers in the system,
Russell's viper, but we find our herbal preparations generally
cope."

The girl transferred her wide-eyed expression from Dr.
Shanthi's face to the band constricting around her arm.

"Then there's that boy," Dr. Shanthi continued. "The one
with the staring eyes. Suffered a cobra bite on his hand yesterday after tea. He's a little shaken but since he's survived so far I
am reasonably confident. Cobras tend to kill quickly or not at
all."

Patrick Russell, who researched the snakes along the Madras
coast in the late 1700s, wrote that the "*Cobra de Cappello*," the
"hooded snake" as the Goa Portuguese called it, "is generally
reckoned of all others, the most deadly." Fayrer agreed. "In case
of a real bite," he wrote, "by which I mean when a healthy and
vigorous cobra...has embedded its fangs and innoculated the
poison, there is very little chance, if any, of saving life, unless
the most immediate and vigorous first aid be given, and even
then..."

The cobra-bite victim sat on his mother's lap, pressing the
unbitten palm against her knee in a manful attempt to keep
himself upright. The bitten hand he left hanging, as if contact
with anything more solid than air was unbearable. As patients
passed him, he withdrew it safely out of reach. To me, this
young boy exemplified a bewildering Indian paradox, one in
which death and life seemed inextricably confused. It was as if
the agents of fertility and corruption had been mistaken for
each other. This boy might survive his cobra bite, but there

were many others who would not. The cobra, which took the
lives of thousands of Indian children every year, was the same
snake that barren mothers prayed to for the gift of children. I
had just seen the snake stones assembled around the pipal tree
in the middle of town, appeals to the snake deities for the lives
of the unborn. I had seen the termite mounds which had gradually turned into shrines. I had heard of the imminent Nag
Panchami, the festival devoted to cobra worship. I knew that
killing a cobra was a significant transgression in India; only a
ceremonial burial could placate the snake's vengeful spirit. I
knew of the cobra's special place among members of the great
Hindu Trinity. Even in this surgery, a room devoted to the treatment of snakebite, Shiva was represented as garlanded with
cobras.

I could certainly appreciate the snake's role in rodent control. It was understandable that Indians appreciate the cobra as
the guardian of the harvest, but did improved grain yield, even
in a largely agrarian community, really count for more than the
daily death threat to family and friends that the cobra constituted? The Hindu take on the cobra was holistic, sophisticated
and redemptive, but its intellectual coherence simply did not
translate into intuitive acceptance in my case. I had been
weaned on the Christian attitude towards the snake, which
seemed philosophically retarded if utterly compelling by comparison, but I could not believe it was this alone that caused me
to think not of bounteous harvests but of Fayrer's dread statistics, the brother Emam's discoloured arm and this young boy
sitting in a surgery, fearful for his life.

Dr. Shanthi beckoned to the Russell's-viper victim, who shuffled towards the table. She bent to examine his swollen ankle,
questioning him briskly as she did so. "The bite site is seeping,"
she told me. "The patient feels faint, weak and cold, he says.
And sick. The venom is currently attacking the lining of the
blood vessels, causing small ruptures and internal hemorrhaging. It's hard to tell how bad this bite is, but he'll certainly have

a bad time of it before the prognosis improves. He can expect to bleed from his eyes and anus before he's over it." She handed him a carefully folded funnel of newspaper containing a powdered, herbal preparation which he upended down his throat. He grimaced and handed over a few grimy banknotes.

"I don't know," murmured Dr. Shanthi, watching her patient limp out into the muggy heat of his uncertain morning. "We may have to give him antivenom."

"The fever belted jungle lands of Borneo, Congo, Brazil, Tanganyika, or any other white man's grave of a tropical world is a pleasant playground for Pollyannas compared to the snake farm of Bombay. Here death summons his coiled battalions every morning before breakfast." It was that man again. Sinclair had become a regular companion in my thoughts. I was thinking of him a few days later as a train carried me through Madras towards Guindy, a suburb on the southern side of the city. As I looked out on lean patches of wasteland interspersed among slum homes, agglomerations of brick and concrete flying patched pennant items of clothing from their washing lines, I remembered how Sinclair had been lent high boots and gauntlets to tour the Bombay snake farm. As I looked down at the Cooum River, glutinous with rank-smelling effluent, where Brahmin cattle nosed at discarded bicycle tires, plastic bags and the stalks of banana clumps, I remembered how he'd watched "turbaned lads" seize "snake after snake, massage him about the neck and drained deadly poison into bottles." And as I glimpsed hoardings for the Star Bone Joint Centre and for Krackjack; the Sweet and Salty Temptation, and one reading "Casio Pager; My Choice—Miss World 1994" (and wondered who ever went to Miss World for pager advice), I thought of the "coal-black cobra" at the snake farm that had reared up and watched Sinclair's legs "with cold, hypnotic eyes."

I had an appointment at the King Institute, the modern, Madras equivalent of Sinclair's Bombay snake-farm experience.

But even I thought Sinclair's "coiled battalions" must be an exaggeration. The King Institute lay in extensive, wooded grounds that were conspicuously silent after the tumult of the crowds and the traffic, and the yellow swarms of two-stroke motor rickshaws. It was approached by silent driveways that multiplied and capillaried to serve the many mildewed outbuildings that attracted a regular flow of brisk-looking people in white coats, but the network of drives seemed even more tortuous than was necessary, as if they had been designed as secret escape routes in the event of some strange catastrophe. They finally converged in front of a large, terracotta building in the high-Victorian style, all crumbling bay windows and sweeping stone staircases, wood panelling and framed portraits of eminent Victorians. It might have housed a clandestine germ warfare project with shadowy links to overseas defense ministries, a gloomy prep school somewhere in Northamptonshire or a lunatic asylum.

The King Institute had in fact been founded in 1898 as a medical research centre and had been producing antivenom for the state of Tamil Nadu since 1980. As arranged, I was met at reception by Mr. Boopalan, the Institute's chief immuno-chemist, a gentle man with a pronounced enthusiasm for statistics.

"Nineteen eighty-six and we were producing just 30,000 vials of antivenom, each of 30 millilitres," said Boopalan. We were sitting in his small office. "This year, we are on line to supply the state's hospitals with over 75,000 such vials. Of course, some bites require no antivenom whatsoever. Others may need as many as 30 vials. Assuming a requirement average of, shall we say 7.5 vials per incident, our production treats 10,000 cases of snakebite annually just in Tamil Nadu."

The great advance in antivenom research had been made back in 1887 when an American professor called Henry Sewall successfully immunized a pigeon against pit-viper venom. By gradually exposing it to tiny doses of the poison, Sewall had encouraged the pigeon's system to develop antibodies that proved highly effective against later, larger doses of venom. The pigeon had acquired resistance. There was, in principle, nothing to prevent those antibodies being harvested, and injected into human victims of snakebite.

What the Sewall breakthrough also established was that the first scientifically acknowledged cure for snake venom derived not from leaves, herbs, poultices, chants or dung but from the snake itself. Just five years earlier, the author Catherine Hopley had written with a striking if unintended prescience that "not altogether discarded...is the ancient belief that in the body of the viper itself is found a specific for its poison." For centuries, the legion folk remedies for snakebite had included eating the flesh of the snake responsible for the bite, or pressing the flesh of the snake against the bite, causing the snake to reabsorb its own venom. The use of snake parts, including sloughed skins, in potions that protected against snakebite was commonplace in Africa. Snake oil was a popular cure for snakebite in the American South and in parts of Europe. But it was the most outlandish claim of all, that a second bite would neutralize the effect of the first, which had now been demonstrated to be closest, in an about-ways fashion, to the truth. A second bite wouldn't neutralize the first but, as Sewall had established, the first bite, if the victim survived it, could help neutralize the second and any subsequent ones.

Over a century later, Sewall's original principle remained uncontested. The process whereby antivenom was produced and administered was essentially two milkings and two injections, in series. The snakes were milked of their venom; the venom was injected into horses (which, it had long been established, were better fitted for the purpose than pigeons). The horses were then milked of their blood and its constituent antibodies, which were then injected into the bitten human.

Boopalan led me down a long corridor that led to the stables.

"Now this is interesting," he exclaimed, stopping to examine a cage resting on a cart at the end of the corridor. The cage contained some forty mice. Apprehensive pink noses sniffed the air, wondering what further humiliation was about to be visited upon them.

"Our product testers," said Boopalan fondly. "We inject each mouse with snake venom. We calculate the LD50, that's short for the lethal venom dose in 50 percent of cases or, to put it another way, the venom dose that causes death in every other mouse. If I were to inject each of these mice with the LD50, exactly half of them should die. The other half would survive; not that they would feel too good! What we do here is inject each mouse with thirty-three times the LD50, a massive envenomation. We then inject them with a dose of our latest batch of antivenom from the horses. The survival rate testifies to the efficacy of the antivenom. I do believe these mice are about to be injected. Let's see how they've got on on our way back, shall we."

We abandoned the mice to the pharmacological horrors awaiting them and stepped outside into the grounds.

"The antivenom we manufacture here is a polyvalent rather than a specific," explained Boopalan. "This is a combination antivenom which is good for treating the four main snakes that Tamil Nadu residents are prone to encounter. These are common krait, *Bungarus caeruleus*, and cobra, *Naja naja*, Russell's viper, *Vipera russelli*, and saw-scaled viper, *Echis carinatus*. We do this because most snakebite victims cannot generally be relied upon to accurately identify the snake that bit them. And administering the wrong specific antivenom can be considerably worse than useless."

We had reached the paddocks. Horses stood in the sunshine, grazing. Boopalan looked on affectionately.

"Excellent condition," he said. "The occasional swelling in the limb. But look at the necks where we inject them. No sign of necrosis or damage to the flesh." Boopalan explained that the Institute had some two hundred horses and mules.

"The army donated many of the mules," he said. "Originally, they were used to shoulder supplies for our units in the Punjab. We even got a few retired runners from the Madras Race Club." The stables, roomy and well-maintained, stood opposite. The impression was of a well-run racing stud. But as Boopalan began to explain what Institute life entailed, I could see that the horses earned their comforts.

"We inject minute doses of all four snake venoms into the horses in a slow-release silicate solution. At the beginning, we inject every month. After three months, the horses begin to receive weekly doses. After six months, they should have developed enough antibodies against the venoms to be considered 'producers' of blood serum, which is when we begin taking their blood. The best horses can produce many litres of such serum in the course of their production lives. From the serum, we remove the impurities. What remains, broadly speaking, is the antivenom."

Mice and horses. Sinclair seemed to have somewhat exaggerated the perils of the snake farm.

"And what about the snakes?" I asked.

"On staff," said Boopalan, "we don't have that many at the moment. We currently hold large stocks of dried venom, you see. But you'll find the few snakes we do have in the pit."

The pit lay shaded by trees. It was circular, perhaps forty feet in diameter, four foot deep and quartered by two cross-hair walls.

"This is where they milk the snakes," said Boopalan, pointing at the few cobras and Russell's vipers decorating the pit floor. "When they have milked them, they release them into a different section of the pit. To prevent the possibility of milking the same snake twice, you see. Now, what say we check on those mice?"

I lingered a moment at the pit, and looked down at the snakes. Death's coiled battalions, high boots and gauntlets indeed. All that nonsense about pleasant playgrounds for Pollyannas. I was beginning to see the advantages of Sinclair's company. Sinclair was a vision of what you might become if you let snakes overrun your imagination. He was ridiculous company, which was perhaps what drew me to him. Nobody was better at encouraging me to come to terms with my fear of snakes than Sinclair, for Sinclair showed me what a fear of snakes could finally do to you: render you a joke, a poor fool.

As anticipated, the mice had been under the needle in our absence. They were largely alive, but they looked none too well for the experience. They had congregated in their distress, as if to find sympathy among each other. Some of them twitched alarmingly, throwing themselves almost clear of the cage floor. Others were almost still, moving only their whiskers. But Boopalan, who was poking the mice through the bars with a ruler to gauge their condition, seemed thrilled by the results.

"Splendid," he exclaimed. "Excellent stuff."

One of the mice looked up at Boopalan, swivelled his whiskers one last time and promptly expired.

16. AMERICA V

The Reverend Jimmy Morrow was standing by his aged purple car outside the courthouse at Newport, Cocke County, Tennessee—where he'd promised to be. And looking much as he'd promised to look: forties, flush-faced but gangly, with that same ruddy cadaverousness, the hale but hauled-in cheeks that were a regular feature of Appalachian people. He was freshly washed, and was made up for church in dark shirt and slacks. He was awkward in his movements; he stooped as if to apologize for his height. He was clearly a shy man, but he welcomed me with a sincere warmth, enclosing my hand in both of his and fixing me with a generous smile. He insisted I park my car and ride in his. A member of his church surrendered the front seat to me and squeezed into the back along with the others.

Jimmy Morrow had the most distinctive Appalachian accent I had ever heard. Following his conversation demanded absolute concentration on the consonants. The vowels were a series of whiny twangs that resembled old-fashioned gunfire, a blazing exchange of Winchesters and Colts. Jimmy was a one-man frontier shoot-out. His utterances ricocheted around the confines of the car. They pinged off the roof and the dashboard, the doors and the headrests. Even members of Jimmy's own congregation could not help ducking at his words.

"So you just come up from Georgee?" he asked.

Alabama, I told him. Which meant I'd driven for over four hours just to be here this evening. Something had clearly drawn me towards another snake-handling service. Perhaps it was the fact that, knowing what to expect, I might understand the

handlers and their motives better this time. But deep down, I
think I just wanted to watch people handle snakes one last
time. I was plain fascinated. Nothing transfixed me like snake-
handling.

"Alabama?" asked Jimmy. Scottsboro, I told him, where I'd
been for a time now.

Jimmy nodded. "Where they had trouble with their rattlesnakes," he replied.

The same might have been said of the Newport area. Cocke
County lies just to the north of the high Appalachians and hard
on the border with North Carolina. In July 1991, just three
months before Scottsboro's Barbee Lane troubles, a man called
Jimmy Ray Williams had died after a bite from a rattlesnake
about fifteen miles north of Newport. Jimmy Ray's father,
Jimmy Ray senior, and Buford Pack, both regular snake-
handlers, had also died a few miles outside Newport, at the
Holiness Church of God in Jesus Name, in Carson Springs, after drinking strychnine back in 1973. Charles Prince, a charismatic local preacher who had slept with rattlesnakes under his
bed as a boy, had died after a bite from a canebrake rattlesnake
at Greeneville, about twenty miles north-east of Newport, in
1985. They said he had gone on handling after the bite, had
Charley. He'd even drunk strychnine. And just two weeks before Prince's death, a North Carolina sheriff had been bitten
while attempting to remove snakes from Prince at an open-air
snake-handling service which the police had adjudged illegal.
And the next town to the east was Parrotsville, the home of
Melissa Brown who had died in Middlesboro, Kentucky, after a
rattlesnake bite just the previous year.

But Newport had more in common with Scottsboro than a
snake-handling notoriety. There was about the place the same
air of steady decline. The same imposing old courthouse
stranded in a town centre that most businesses, the motels, the
retailers and the restaurant trade, had long since forsaken for
new pitches out on the highways and in the new malls where

they had all the space and the parking lots they could use,
which meant the opportunity to do some volume business for
once. They had left behind turn-of-the-century, dark brick
buildings, handsome things with boarded-up doors and windows, and walls that attracted a patchwork of bill stickers and
some unmemorable graffiti. Lured by the give-away rents, a few
optimistic souls had turned such premises into cosy, like-it-
used-to-be coffee shops of the sort they'd seen in the heritage
quarters of Chattanooga or Asheville, with chocolate brownies
on cake stands and homey chipped china, fliers for local events,
views of the railway line and copies of the local newspaper, the
Newport Plain Talk, about the only business to have kept faith
with the downtown district.

Jimmy's church was way out of town, but then such churches
always had been. The Holiness movement had its origins deep
in the countryside. The way the world was these days, they
might have to do their working in towns since town was the
only place they could hope to find positions. Might even have
to live there, but worshipping, they would do that in the woods
as they had always done. Our journey had a purpose, then,
other than arriving: it put a distance between Jimmy's church-
goers and what they perjoratively called "the world." Town was
where you found "the world," whether you looked for it in the
bars or in the movie houses or in the homes of other men,
whose wives knew just how long you had before they could expect their husbands home.

Newport had always had a reputation for worldliness. In the
early 1800s, a traveller called Henry Ker had described it as "the
most licentious place in the State of Tennessee, containing
about twenty houses of sloth, indolence and dissipation." More
moonshine, it was said, had come out of Cocke County in its
time than anywhere else in the region.

So that was it. It was not so much a lift we were being
treated to as a return to Holiness. We were being carried back
to Christian purity in an old purple car heading east and south.

And sure enough, once we had cleared the Newport city limits, the landscape began responding to the notion. In the sunshine of a spring Sunday evening, the Pigeon River valley was breathy with rural innocence. Pastureland ran down to the river; brown cows floated legless in a rising mist; red barns evoked white-haired grandfathers, benevolent in faded blue overalls. I looked up from the window into the beech and birch trees that canopied the road. Their new foliage, translucent in the sunlight, threw soft shadows that streamed past the car, dappling the road in our wake.

"Beautiful, ain't it?" shouted Jimmy Morrow above the rattles in his engine.

Jimmy had been pastoring in the Holiness way for eight years now. His first church had been near his family home out at Del-Rio, on the French Broad River to the east of town as you climbed into the foothills. In 1993, the members had built themselves a new church to the south of Newport, this time in the valley of the Pigeon River. They had chosen a place called Edwina. Edwyna, as they pronounced it, was a collection of a few shacks and chalets in the woods, all that remained of the community which had once served the hard-by Tennessee and North Carolina Railroad.

"We built the church with our own hands, me and some of the members," explained Jimmy. "We wanted somewhere isolated, off of the road, like where we've always been."

The people of Edwina, a seasonal lot, mostly hunters, had been unenthusiastic about the idea. Building the church in 1993 had earned several members some pretty lurid threats. One afternoon, Jimmy had even taken a beating.

"There's people out there don't like us serpent handlers," he explained. "Think us crazies, Holy Rollers or something like that." He spoke without rancour; Jimmy was quite used to the things people said, calling him and his kind barefooted, backwoods people, hillbilly types who had no business in the twentieth century. And he had been wise to the things they did, letting down tires and stuff, until that afternoon in 1993 which was the first time it had ever got physical.

"Those people should know their Bible," was all Jimmy said. "Then they would know that serpent handling is a direct command of Jesus Christ. All we're doing is obeying the Word."

We left the tarmac for a forestry track which wound deep into the woods. Edwina's few properties appeared among the pine, birch and elm trees which sliced the late sun into slanted stripes. Jimmy's church stood close to the end of the track against a backdrop of woods. It was a simple grey block building, fifty foot long, with a pitched corrugated iron roof and high windows punched along each wall. Beneath the front porch, steps ran up to double wooden doors. To the left of the doors, a neatly painted board read The Church of God in Jesus Christ's Name. The board stood atop a plinth which was crazy-paved in a rare, decorative concession. Inside, the walls were lined with chipboard. Grey-painted pews six rows deep gave way to a raised dais, and, in the middle, a wooden altar on which stood a small wooden cross.

Soon enough, Jimmy's gathering was complete. There were just twelve adults and a few children, but at least the guitarist had made it. And somebody had brought the cymbals.

"High times," Jimmy told me, "we had forty regular members. Now, maybe there's twenty regulars always come 'cept when they got work, when they can't." Jimmy thought about their absence for a moment before judging the excuse acceptable.

"A man has to work to live, don't he?" he said, and so got on with it.

The gathering started singing. They were stately, loud and even robust but they did not have the numbers to generate an atmosphere like Carl Porter's lot in Georgia had achieved. The rafters didn't shake at Edwina, not like they had at Kingston, alarming even passing truck drivers with the cacophony, but they hummed with a reassuring communal intimacy that began to draw me in. I had always imagined the snake-handlers as freakos, types I might observe with curiosity, mild distaste and mostly detachment, as if their lives meant nothing to me. But the mood in the church was infectious and inclusive; I was not here for long, it seemed to say, but so long as I was I might sing with them. So I joined them where I could. I think I even made a pretty good go at "Amazing Grace."

Jimmy set to preaching when the singing ended. For a while, he just looked absent-minded, prowling round the altar and patting pockets like he'd mislaid something, saying nothing. But he was into his stride the moment he began. The cadence of his words immediately sounded familiar; just, it eventually struck me, one I'd heard hitched to a different vocabulary. Jimmy was talking just as auctioneers did. It was the same rapid rattle of words in which only phrases survived, and the same pauses for breath where Jimmy inserted his most conspicuous preacher's tic; a rising *whoop!* that suggested imminent loss of control followed closely by a blissful, purred *hallelujah!*, a final reassurance.

"You know the Bible says without Holiness," said Jimmy, "no one shall see God, for we're living in a wicked world...And you know, the Bible says there's another place...And the Bible said he called them to go preach the gospel, and I'm telling you children, whoop! hallelujah! after he was resurrected, he told his disciples to go out into the world, and these should be the signs that followed them...and whoop! hallelujah! you know how the Bible said Heaven and Earth shall pass away...and in Timothy 3 the apostle Paul says that in the last days there would be perilous times and that's what we got right now whoops! hallelujah!" It was a remarkable expression of faith, but it was no more than a litany of Jimmy's favourite Biblical references, as if he wasn't fool enough to try and improve on the Lord's message.

As he preached, Jimmy calmly opened the wooden box that lay upon the altar and removed the two young rattlesnakes that were inside it, the evening's only snakes. He seemed reflective as he arranged the snakes, patterned like autumn leaves, in a loose plait in the palms of his hands and supported them above the altar in a slant of late sunshine, an offering devoid of theatre or pretense. When Jimmy passed the snakes among the few men who had gathered to take them, he seemed to have simultaneously handed them the gift of his certainty, and it was as if a magical communion, an act of intimacy was taking place in the small church which beckoned me, and for a moment I wanted to move forward.

Later, a young man with rigorously brushed hair stood up to testify. He said how he had once been impressed by the spirit of faith he had witnessed at a Holiness service just like today's. At the end of that service he only knew, he told us, that he wanted the same certainty.

"I prayed," he said, "and I fasted. I talked to God and I said, 'Lord, what must I do, what must I be, to be like those people? If I could just feel like they feel for one time, Lord, that would be beautiful from where I was standing.' Brother Jimmy, you know I laid down on my bed one night and I was lying there and I was so tired and I remember a voice spoke unto me, Brother, and it was not a voice of the spirit but it spoke unto me as a man would talk. And the voice said, 'When your time comes, it said, this is what you're going to feel.'

"Brother, I never felt nothing in my life that felt so good. It began to run all over me and I began to melt and run down the bed and everything. When I come to myself the next morning I jump up and said, 'Lord, I went to sleep when you was talking to me.' But the voice that came back was the Devil, and the Devil said, 'That wasn't the Lord talking to you, you dreamed every bit of that. The Lord can't give you that feeling.' I thought about that and I said, 'Devil, if I had dreamed that, then I can dream that feeling back by myself,' and I worked all week long to reach for that feeling without the help of the Lord. But that feeling wasn't sent from me, it was sent from heaven and I couldn't get it on my own. I could only get it from the Lord. And I thank God that one sunny day you people showed me some serpents and that's where I began to feel that feeling, that same feeling I felt just now when we was handling.

"And Brother Jimmy, I'm here to tell you that you'll do anything to get that feeling in your life. You'll fast, you'll starve yourself to death, you'll move out of the house, you'll sleep in the woods, you'll never look at another dish of food until you get that feeling back. If you've ever felt that feeling, you'll never go no other way."

And Brother Jimmy smiled at the young man as if these were things he knew only too well.

At the end of the service, Jimmy locked up, put his rattler box in the car trunk and drove us back to town. On the way, I asked Jimmy how he could be so sure that the end was near. I'd heard others in the Holiness community like Patricia Hale down in Georgia saying the same things.

"Well, we're living in a time where everything's breaking down," he reasoned. "It says in Matthew 24 how in the latter days there would be false prophets, and famines and earthquakes and all this would come up on the face of the earth. We never thought there would be terrorist acts, you know, like in Oklahoma where they blowed up the federal building. We never thought there'd be babies murdered in their mothers' wombs, like what the abortionists are doing. Everywhere are signs that these are the latter days."

I sensed a ruefulness pervade the car. Jimmy's certain knowledge of the apocalypse seemed to have ushered in a collective gloom, as if there might not be many more such occasions as this, driving through a warm spring night in the Appalachian foothills, with moonlight flickering along the Pigeon River and the chorus of frogs off the water meadows. When I turned, however, I realized that I had misinterpreted the mood. Two of the church members were asleep, the others were staring from their respective windows. All four were smiling contentedly. For them, I realized, the end of the world was just the beginning. The calling home.

~

Glenn was right about the snake. This time, it needed no encouragement to bite.

Within minutes, Darlene was in serious trouble. She had to work hard, drawing deep to fill her lungs as if she had lost the breathing instinct. Her body was tingling. It had felt strange for twenty-four hours but nothing like this, a startling sensation sudden with pain that began in her hand and travelled up her arm in topsy flutters, standing matchsticks up under her skin, turning her hot and cold.

In his strange way, Glenn had become attentive again. He led her back to the house and lay her on the couch. He even brought her the glass of water that she cried out for. But she threw it up immediately. The vomit splattered the side of the couch and puddled onto the floor. Glenn disappeared, mentioning something about the bucket and mop, but soon returned empty-handed. He leaned in the doorway. He was thoughtful.

"Now with you trying to commit suicide," he said, wagging a smart finger at her, "why would you have a bucket, you know?"

Even as she blacked out, it occurred to Darlene that Glenn had got it wrong and she wondered where he'd got the suicide notion from. The next thing she heard was Glenn's voice, as if through a thick wall, asking her whether she wanted to go to the hospital. She seemed to have forgotten that she'd heard that one before. That he'd promised to take her to the hospital only that morning. All she wanted was for the pain to pass. And hospital was where they could do that for her. She managed a nod.

"So all you gotta do, Darlene, is write these little notes for me," said Glenn. He hauled her upright on the couch and passed her a pen and paper.

"I gotta do what with these?" Darlene asked. Never before had he asked her to do writing. Her head was spinning.

"You just gotta write what I say, Darlene. Then you get to go to the hospital." So Darlene wrote. Not so good, but she just wrote what Glenn said—and what with the drink, Glenn didn't manage that so well either. "Marty," she wrote, "Marty, I love you. Do what Daddy say. Don't blame Daddy because he loves me and trying to help me. Daddy is asleep and he don't know what I'm doing. I don't want no help and you and Daddy live right, OK?"

"That OK?" mumbled Darlene. Glenn was reading the note.

"That's OK. Now you gonna write me another one."

Darlene wrote. She simply did what was asked of her. That way, the sickness might stop. Later she would understand that she'd been writing her own suicide notes. Glenn liked to believe that, what with Cleopatra killing herself with snakes and Darlene's snake-handling reputation and now these notes, he had all the makings of a solid defense. When Darlene finally finished writing, Glenn removed the second note from her, nodded, then lay down on the couch opposite and told her not to move. Otherwise, he added, he would blow her brains out with the gun. He was finished; those were the last words he ever expected to say to her. He tucked the gun under the pillow, closed his eyes and settled down for the night.

"But you said you'd take me to the hospital." Glenn didn't reply. The drink was bludgeoning him into sleep. Darlene began to panic. Some instinct, long overdue, told her that she was dying. She screamed at Glenn, but his only reaction was to knit his eyebrows at the din. Then his face relaxed, his breathing evened out and he began to snore, and Darlene realized with a shock that he expected to find her dead when he next awoke, when he might have a morning hangover but at least she would be gone from his life. He'd got her bit twice, lied to her about the hospital twice and threatened to blow her brains out any number of times. Then he'd settled down for the night and left her forever, covered with vomit and whimpering with the worst pain she'd ever known, with a final raised eyebrow to let her know what he thought of her.

A surge of anger like she'd never known hit Darlene. She suddenly determined that whatever else she was going to do in her life, she would make sure that Glenn woke up to more than a hangover. Glenn was going to lose his corpse. He wasn't going to get his fresh little girl, whoever the bitch was, just a whole lot of trouble. Darlene didn't want to die. She didn't want her shabby life over with. And at that moment, she began fighting for her life. But she couldn't call out to anybody. She was going to have to do it all on her own. The thirst was terrible. The prospect of movement exhausted her. Her eyesight was going. The room seemed full of water. The furniture was swimming. She slid off the couch and began crawling towards the kitchen. The pain in her arms made her whimper, but she no longer permitted herself to cry out. The point was now not to disturb the man that she'd been busting to wake moments before, and she bit her lip against the pain.

Then she noticed the telephone. It was on the table, which wasn't usual. She remembered how Glenn had unhooked it. He hadn't wanted to be disturbed. Darlene dragged the phone into the kitchen. Her mind was spinning, tumbling her over in swells of faintness so all she could do was cling to consciousness until each wave had crested and passed. Then, in the brief moments of clarity, she checked for Glenn's snoring and set to work hooking up the phone. She remembered where the phone socket must be. When the phone jack slid into place, she grabbed the receiver; she had a line. She had no trouble remembering her sister's number; she had called it more times than she knew. What was hard was hitting the right digit buttons. But the voice that answered after a few rings was Debbie's.

"Yep," Debbie said, her voice guarded at the lateness of the hour.

"Listen, Debbie," Darlene whispered. "Just listen. Glenn got me bit by his snakes. He's trying to kill me. I need an ambulance. I need to get to the hospital quick. He's sleeping now, but he's got the gun. Tell them I'll meet the ambulance out on the lane. Out by the Chambliss place. Tell them to keep it quiet. No questions. Not now. Hurry, Debbie."

She put the phone down. She retched. Then she was on her hands and knees, moving towards the back door. It swam away and left her, but slowly it came back towards her. She reached up and found the handle.

~

I had a flight home from Atlanta the next evening. Since it was a beautiful morning, I thought I'd spend the day touring the backroads. And if time started pressing, I could always pick up Interstate 75 near Chattanooga and hurry south.

So I headed west out of Newport. The jade parapet of the Appalachians kept pace with me to the south as I passed through wooded hollows and farmland. There were tractors in the fields, country music on the radio, and those roadside signs that were always hemorrhaging their letters to keep me amused. One, outside a country gas station, said W GNME. Wheel alignment, I shouted. Another, more difficult, said UNTI IES. For a moment I wondered why a gun shop should advertise "Auntie's Pies" before "Hunting Supplies" came to me and I slapped the dash in triumph.

It must have been somewhere around Sevierville that things began to change. For one thing, the road suddenly became Dolly Parton Parkway and the likes of Dolly Parton did not usually give their names to the local roads, generally called things like Carson Springs or Stoney Fork, Grasshopper, Deep Creek or Pigeon Valley. Then there were the billboards which began as occasional stands but soon burgeoned into thickets. I'd left the old signs behind, the ones that offered, once decoded, to patch up your car or supply you with fishing bait, firewood, gun cartridges or fresh vegetables. Now I was being entreated to visit Dollywood; The Smokies Entertainment Capital, Carbo's Police Museum; As seen on the David Letterman Show, and Pop's Catfish Shack. The weirdest thing; I had stumbled across modern America.

The cause of all the fuss was the nearby town of Pigeon Forge where, it transpired, Dolly Parton had been born and bred. By and by, Dolly's Tennessee Mountain Home had become a resort town with a thriving music scene. So I might have visited Pigeon Forge for the Louise Mandrell Theater ("The music will charm you…the magic will enchant you…the moment will touch you…"). I could have stopped at the Dixie Stampede ("horses, beautiful belles and tough-as-nail heroes"). I could have visited the Ogle's Water Park or the Elvis Museum ("his last limousine among the many items displayed") or the Smoky Mountain Car Museum. But what decided me was the sign that read "Bible Factory Outlet. 1000s of Bibles at Half Price."

The traffic had thickened to a slurry by the time I reached the town's several-mile strip of banks, chapels, budget inns, realty offices, show venues, Burger Kings and Taco Bells, grills, collectibles boutiques, factory outlet stores and helicopter rides to the mountains (perhaps to sample that 2,000 mile footpath, the Appalachian Trail, which passed just above the town). The signs to the half-price Bibles never let me down. I turned left at stoplight 2, as directed, and turned into a large car park. I drew up near a sign that told me I had missed, by a few weeks, Pigeon Forge's Cutest Little Baby Face Contest, and glowed inwardly at my timing. A flat-roofed arcade of concrete shops coralled the car park. On one, BIBLE FACTORY OUTLET was written in blue. I stepped inside.

They were right about the Bibles. You could have your Bible any way you liked it. You could even get the New International Version on audio cassettes, forty-eight of them, for just $89.99. There were also mugs and cards printed up with biblical

proverbs. There were ornaments including little angels—
"to watch over you and your loved ones as you sleep." There
were crucifix necklaces, and rings and bracelets engraved with
Christian messages. And there were the T-shirts—one depicted
a Hell's Angel on a Harley above the legend "Take a Second
Look at What you've Passed." And in the biker's wing mirror—
Calvary, with Jesus nailed to the cross above the words "Objects
in Mirror are Closer than they Appear." There were wall charts
showing the biblical history of the Middle East. "Fascinating
Facts," the chart promised. "Noah's Ark may have landed in
Turkey. Daniel was a POW in Iraq."

I left town pretty much immediately. I didn't stop for any of
the attractions, not Patriot Park ("see an actual Patriot Mis-
sile"), not the Mountain Valley Chapel ("Marry on a Mountain!
...no blood test or waiting period"), nor even a round of golf at
Bunny Land Mini Golf ("Play mini-golf with live bunnies").
Soon, Pigeon Forge was a welcome blur in my rear-view mirror and a few final signs looking forward to my next visit. And
as I picked up the road that led to Maryville and thence to
Cleveland and into the wide flow of Interstate 75 South, I wondered who were truly the crazies in these parts.

17. AUSTRALIA V

Taipans are common in the Mareeba area.

—*Cairns Post*, September 5, 1991

OCTOBER 1996

"Why don't I show you where it happened?" Clive Brady suggested.

We were sitting in the front room of Brady's Mareeba home, a modest bungalow on Middlemiss Street. Middlemiss Street had a retirement air about it. The poinciana trees were regularly spaced and heavy with bean pods that hung like burnt-brown crescent moons. The shurbs in the well-tended front gardens were garish, spiky and vaguely extraterrestrial in the tropical manner, and everywhere was the smell of freshly mown grass. Cars were neatly positioned under Plexiglas ports where the sun flared.

Clive was dressed for the tropical summer; he wore plastic thongs and a faded swimming suit, and was bare-chested. Despite a cheery, neighbourly manner he had wary, hooded eyes and a hairstyle which further contributed to a strangely train-robberish effect; what remained of his greying hair had been cropped close to his skull. The skin around his neck was tortoise-folded from the sun. Clive had lived in Mareeba almost all his sixty-nine years.

It had happened almost exactly five years before. Two weeks before his sixty-fifth birthday, two years before he would retire from the railways where he had worked since he was fourteen, and although he did not know it, just a month before Darlene

Collins would get bit in Alabama, U.S.A. I wondered how Clive felt about revisiting the spot. I confess I was apprehensive myself.

"I was a bit uneasy at the beginning," Clive admitted. "Nerves were shot, you see. Slightest thing made me jump. I must have stayed away for a good month after it happened. Now, of course, I'm always taking a cool-off in the river. I keep my eyes open, mind, not straight in without a care like I used to. And I certainly don't bother with the pumpkins any more."

We stepped out into the sunshine and Clive led me down the final stretch of sharply sloping Middlemiss Street. At the bottom, the street terminated in a circular turning space, giving way to dusty eucalypts that attracted occasional mobs of frantic yellow lorikeets, and a palpable wilderness sense beyond. It was here, where Mareeba came to an end, that a track turned left and led down towards the Barron River.

A coach change on the track inland from Port Douglas to the tin mines at Herberton, Mareeba was settled in the late nineteenth century. The Barron meets Emerald Creek at Mareeba, whose aboriginal name means "Meeting of the Waters," before eventually finding its way into the sea north of Cairns. The town had been arranged in a grid, but one that was nibbled and irregular at its eastern extent where Middlemiss and a few parallel streets gave out at the meandering Barron. These roads ran to within feet of the river's bank, as if to seek out its promises of shade and cool water. Which was pretty much why Clive Brady had been making regular journeys down there, after school and then during his lunch-breaks from the railway sidings, for over sixty years. He knew the path we were walking as well as anybody.

Seeing to the rail freight that passed this way all those years—cattle mostly but also some tobacco and corn—had kept Clive fit. In his time, he'd uncoupled more goods wagons than he cared to remember. Leaping from wagons, day in day out, had finally done for his knee, mind. Gone gammy, it had. Still, as the doctors had said, it was his general fitness that had stood him in good stead that day. Had probably kept him alive.

Clive was born on Middlemiss Street, in the house next door, in 1927. The Bradys had been here since the early days. They even had a nearby municipal park named after them. Clive was a reminder of how recently Mareeba had been settled; the generation before him had watched their fathers hack the town from the bush. They had then had the town's parks and streets named after them. And their descendants hadn't yet had time to disperse. As the phone book showed, there were seven Hastie entries for Mareeba—Hastie, I discovered, had been the contractor responsible for several of the town's first roads—and two of them lived, stylishly enough, on Hastie Street. There was a Courtney Street (the town's first station master), a Rankin Street (the town surveyor) and a Pares Street (the auctioneer), with descendants to match. Middlemiss had been the town butcher; two Middlemiss families still lived in Mareeba. Constance Street and Hort Street apparently commemorated the Christian name and surname respectively of the same woman, suggesting some local consequence. That there were no surviving Horts led me to imagine Constance Hort as a Queensland spinster of the old school; inclined to hauteur, a touch eccentric, the last of her line.

Mareeba had emerged, it seemed, as a strikingly schizophrenic town. Caught between contrasting impulses, Wild Frontier on the one hand, Cosy Provincial on the other, its townsfolk could not decide whether to trash the pubs or to organize charity evenings, to yell *yee-hah!* (the town had a famed annual rodeo) or to call *three spades* on genteel verandahs. Reading the silver and gold mining lease applications in the local *Tablelands Advertiser*, staked claims from another time, was liable to inspire a rush of Klondike enthusiasm so you ended up shelling out on a pickaxe and a supply of beers, and heading for the dry creek beds before you knew what you were doing. "Commencing at a datum post," they typically read, "marked

E. H. Jennings 23/9/96, situated in the northwestern corner, 275 degrees 0 minutes and four metres from blazed tree at the junction of two unnamed gullies approx 1200 metres north east of Spear Creek thence by 23 degrees 0 minutes for a distance of 303 metres thence by 40 degrees 0 minutes..." Then you might turn the page to read of coffee mornings at the Rotary Club, and breathless talk of prize vegetables, and you might think less of monster strikes and mother lodes.

Then again, the town could be rowdy. Concealing themselves in roadside mango trees from where they lassoed passing motorists in their open-top cars, regularly causing serious injury, was said to pass for fun among the local lads in the town's early days. The place swarmed with hotels: Eccles, Lloyd and Strattmann Streets all took their names from early hoteliers. Above the door of my hotel, I had noticed the board stating the serving of spirituous liquors etc. to be the business of the establishment was headed *Primary* Purpose. Which conveyed the tacit acceptance that, OK, hell and the rest of it as well had also been known to occur on the premises.

Mareeba even seemed confused about its location. It had learned not to look for company at most points of the compass; go north 500 miles to the Torres Strait at the top of Australia, and you'd soon stop making cheap jibes at the expense of Mount Molloy, just about the biggest settlement you'd meet on the way. Go west, and you'd hit nothing bigger than an outhouse before the distant Gulf of Carpentaria. But go east and Cairns was only forty miles away—a big city on the doorstep by Australian standards—and it was from Cairns's proximity that Mareeba seemed to have drawn a surprising civic confidence.

The centre of town lay just a few hundred yards west of the Barron. Here were the stockyards and the railway line, the local government building, the hotels and the tractor-part suppliers, the radio station and even a mosque. The feel was often old-fashioned. Flat-roofed buildings in sun-faded pastels housed the likes of Enzo's Hot Bread Drive-In Bakery, Hair by Garry

and the headquarters of the Tobacco Leaf Association of
Queensland (where "leaf" truly seemed the clincher). The
Capitol Arcade, announced by distinctive fifties signage—each
letter assigned its own foot-high plastic plaque that was shaped
like a blunt, upside-down arrowhead—was full of milk bars and
knitting shops. But the town also boasted an attractive Heritage
Museum and excellent sporting facilities including a swimming
pool and a golf course, a rugby pitch and squash courts, tennis
courts and soccer pitches. It even had a racecourse. Which
Clive might have visited that day, and so spared himself a visit
to the hospital. Mareeba had one of those too. It happened to
stand just above the stretch of the Barron where Clive liked to
swim, a fact he would come to appreciate.

That morning, I had found Clive kneeling by his front step,
spraying clouds of insecticide beneath the carpet.

"Infestation of bloody ants," he told me.

That seemed about right; ants, an outsider might guess, was
about as bad as Mareeba got. Clive knew differently, and had
known so since his childhood.

"There were always snakes around," he told me. "Bags of
them." As a child, back in the 1930s, he had lost an uncle to
snakebite. It had happened on the banks of the Barron, some
miles upstream near the dam at Tinaroo.

"Mum said Uncle Jack was always chasing possum," Clive
told me. "Up to here in scrub too often," he said, pointing at his
chest. "They took him to the hospital in Atherton but he was
dead by the time mum got to him. Tongue swollen, terrible
bleeding and vomiting—like what happened to me."

Another Mareeba man, Vic Sibley, had also died from snakebite received on the same stretch of the Barron during the
1930s. Sibley, a stock inspector, had been on a fishing trip with
some mates. At eight in the morning, one of them heard Sibley
call out that he had been bitten by a brown-coloured snake.
They helped him across the river to a waiting farm lorry that

one of the party had gone ahead to alert. The Atherton ambulance met the group at Tolga but Sibley was, in the words of the doctor at Atherton Hospital, "already cyanosed and comatose, with paralysis of the respiratory and throat muscles" upon admission. His wife arrived an hour later to witness his dying moments. According to the *Cairns Post*, Sibley "enjoyed wide popularity," and "his tragic ending cast quite a gloom over the district."

Nor were humans the only victims of snakebite. According to Clive, Mareeba stock and horses were regularly found dead. You could tell it was snakebite from the blood at their nostrils. The Bradys were horse people. They loved the races; one of the Brady boys was even a jockey. They knew about breaking horses too. Clive remembered how he had had some horses paddocked on the other bank of the Barron, not far out of town back in the early eighties. He was leading one from the paddock one day. There had been heavy rain; the Barron had been up and had flattened a couple of sections of the fenceline that ran down to the river, the rip current combing it hard against the ground. As they crossed over the flattened fence, the horse appeared to flinch. Clive only assumed the animal had been nicked by a loose strand of barbed wire, but ten minutes later the horse was in obvious distress. When they reached the stables, Clive got his son to help him unsaddle it, thinking to get the horse straight into the shade of a nearby gum tree. But the moment the horse was free of its saddle and bridle, it just staggered around in a series of broken pirouettes. Like it had gone spastic, said Clive.

"His eyes were staring mad, the way they said mine went some ten years later. Sweat was coming off him in buckets. Literally, buckets. And there was blood at his nostrils. It was like he just couldn't breathe. And, you know, that beautiful horse just dropped dead."

Soon, Clive's brother appeared at the stables. He'd been out repairing the fenceline and had just killed a taipan up there, he told Clive. A seven-footer. I was beginning to understand why country Queenslanders knew the taipan as the horse snake.

It was a Saturday, the last day of August, 1991. Clive had spent the morning mucking about at home. He'd cut the lawn. Done a few of the chores that needed doing. Was taking it pretty easy. Could have gone to the races with Blanche but Clive had decided not to bother. He had had it with three-horse races and old nags every one of them, which was all the Mareeba course seemed to attract these days. The wife could go alone.

The path was levelling out onto a flood plain. The river came into view. It was low, perhaps twenty-five feet between the banks. An overhanging canopy of foliage dappled it with shade. There were patches of clear where you could see the sandy bottom. Elsewhere, emerald weed swirled gently. It was a beautiful river, a river lifted from some temperate willowy idyll, but the sight of it made me think of Clive's uncle and of Vic Sibley, and it made me shiver. Clive motioned to the riverbank where it ramped gently into the water.

"That's where I got in that Saturday," he said. "The same place I always get in." He would tread water as he always did, his arms gently flapping, and let the current carry him, feet first, a few hundred feet downstream. From the riverbank, we began tracing Clive's progress downstream that day. The walking was easy. There was thin grass and minimum leaf mould so I could see where I was putting my feet. Down by the Barron, I liked to see where I was putting my feet.

"The river was running gentle," Clive remembered. He stopped on a grassy bluff above the bank. "It was here I heard noises," he explained. "A family picnicking on the bank above me. I just waved vaguely and floated on."

About a hundred yards further on, Clive had had enough swimming. The nearside bank at that point was a bit of a scramble, so he clambered out, as he routinely did, on the easier far bank and began walking upstream to a crossing place he knew. On the way, Clive idly noticed horses had been down there; they'd left hoofmarks all along the bank. Clive was keeping his eye on the ground that day, looking for wild pumpkins either side of the path. After the rains, he used to throw pumpkin seeds so there might be a few to collect later in the year. That day Clive thought Blanche might like a few pumpkins; the gesture might cheer her up, especially if she'd lost on the horses. Upstream, he could hear a dog barking. *Brian Eakins's lot*, he suddenly realized; they were the picnickers he'd passed. Brian and Clive worked together on the railway.

Clive stopped, took my arm and pointed across the river.

"Just there," he said. "That's where it happened."

It was an unremarkable patch of ground, all dried mud and a thin growth of weeds through which a track of sorts could be traced. Below it ran the river; above it, bush gave way to steeply sloping paddock land. It was a place you'd have thought nothing of. Just as Clive was thinking of other things as he passed that Saturday five years ago. For a few seconds, a string of associated distractions had occupied Clive's mind; the hoofmarks of horses watering, his wife watching horses racing, the ripening pumpkins that she might appreciate, a dog barking and a family he suddenly recognized as friends, and idle thoughts of work come Monday. For a few seconds, these thoughts had carried him forward and kept his glance from the track ahead of him. It was as if he had temporarily relinquished the right of way. So it was that when Clive next looked ahead of him, a large snake had taken possession of the track in his absence and was moving straight towards him. It was a taipan.

"Thinner round the neck than other snakes, they are," Clive told me, staring grimly across the river. "Which gives them a distinct head. Cream coloured around the mouth and the eyes. Orange, the eyes, plus something about the way the eyes are set; evil-looking buggers."

The taipan was about six foot away. Its rolling, copper-coloured body was travelling fast, its head clear of the ground. "It was a six-footer," said Clive. "And it just looked like it meant trouble."

He tried to get clear, moving to jump down into the thicker weeds just above the water's edge. But he was sixty-four and his knee was gammy, and he saw the snake "a yellowish flash," coming at him from the corner of his eye, even as he began to move. For a split second, he knew he was about to get bitten and he guessed dying must feel like this, and he wished he'd gone to the races. With an elastic uncoiling, the snake flew at Clive, striking him on the left knee and knocking him off his feet. He grunted and landed heavily. From the ground, he watched the snake disappear into the scrub.

Everything had gone quiet. It wasn't supposed to happen like that, thought Clive. OK, taipans were known to be aggressive at this time of year, but they weren't meant to make for you plain as day, arrowing across open ground like this one had. He looked down at the puncture marks just below his knee, several tiny pairs of blood spots, some of them smeared now. Later, they would tell him that the taipan had struck him seven times. Clive guessed that he had just taken a massive hit of taipan venom. A taipan's average venom yield is sufficient to kill 12,000 guinea pigs. Clive was no guinea pig. But nor was he equal to 12,000 of them. He must stay calm, he told himself, but he must also get help. He remembered the Eakins family and guessed they were no more than a hundred yards upstream, on the far bank. He got to his feet and began walking towards them. He could already feel a fiery sensation around the bite site. It was not that painful but it told him the venom had gone to work and he picked up his pace. What scared him was what happened next. At first he thought he must have something in his eye, a smear of mud or a bit of leaf mould he'd picked up from the fall perhaps. It was when he stopped to wipe his eye that he realized his eyesight was failing. The focus had gone. There was just a blur, bright towards the sky and dark at his feet, sand-wiching the suggestion of trees, billowing trees as if drawn by a child. Clive now began to think he might have imagined the Eakins picnic and he sank to his knees, submerged in panic.

"Help," he cried out. "I just got bit by a snake."

The taipan's traditional prey, rats and other rodents, tend to experience next to nothing before nothingness, such is the overwhelming effect of the snake's venom. On humans, of course, it acts rather more slowly, which is instructive if un-pleasant. As Eigenberger had discovered, the human victim ex-periences the venom's stage-by-stage subtleties emerging, the symptomatic developments, as if in slow motion. Not that it felt slow to Clive. His failing eyesight was the first sign that fast-acting components in the venom, the vanguard in the taipan's armoury, were successfully targeting his nervous system. Neu-rotoxins were going straight for the neuromuscular junction, the engine-room interface between nerves and muscle, where they set about closing down the functions of the nervous sys-tem. Slurred speech, difficulties swallowing and other alarming symptoms would follow. But it was the venom's effect on the respiratory system, paralyzing it and so preventing Clive from breathing, that constituted the initial threat to his life.

The call for help rang out across the river, shattering the Saturday calm of the Eakins picnic. Brian leapt to his feet and waded across the river with his teenage son, Rob. They reached the stricken man in seconds.

"Strewth, it's you, Clive!" Brian exclaimed. His workmate was breathing heavily and was drenched in sweat. His face had turned the colour of ivory.

"Taipan," Clive said. Brian laid his hand on Clive's arm, told Rob to stay with him and waded back across the river to get help.

The bite could not have been worse but nor could the hospi-tal have been nearer. Drenched, Brian ran up the track and burst into the hospital reception. There had been a snakebite down by the river, he shouted. Assistant nurse Bernie Tonon grabbed a compression bandage and followed Brian down the track while the receptionist rang for an ambulance.

They reached Clive in minutes, and soon had his leg in a pressure bandage, the standard treatment for snakebite. But the venom had had ten minutes unrestricted access to his system, and was already wreaking general traumatic havoc. Cramps were racking Clive's stomach. He felt nauseous and weak. His speech had been reduced to a delirium. His eyes were unfocused. Saliva dribbled from his mouth.

By now, another specific venom component had gone to work. Clive's muscles were being attacked. They were literally being broken up and scraps were being cast adrift on the bloodstream. In time, this muscle shrapnel would reach the kidneys and so clog them, causing renal failure. Simultaneously, the neurotoxins had made serious advances on Clive's nervous system. He could hardly move now. They stretchered him across the river towards the waiting ambulance. But as they carried him up the far bank Clive stopped breathing. His lips had turned a lovely, duck-egg blue.

The venom had knocked out the respiratory system. Clive was now suffocating. They scrabbled to the top of the bank and laid down the stretcher so Bernie could administer artificial respiration. She pumped the old man's chest. Then she gave him mouth to mouth. He convulsed, vomited into her mouth and then began breathing again.

He was at the hospital three minutes later. His blood pressure—80 over 60—was dangerously low. His heart was hardly beating. He needed antivenom immediately. With only his word to go by, they could not be sure that Clive had correctly identified the snake. Since the wrong antivenom could kill him on its own, they gave him a polyvalent.

Clive came round soon after and began vomiting.

"I spewed and spewed," he remembered. "It was like I was dragging stuff up from my boots. It was the vilest stuff you could ever smell or taste, like horse piss. I really thought my gut had turned right over." By now, the news had reached Blanche at the racecourse. Within minutes, she was drawing up on the hospital forecourt. Here, an inhuman noise, a hollow racking, retching sound like nothing she had heard before, reached her from within the hospital.

At around six P.M., they drove Clive down to the Base Hospital at Cairns where tests confirmed taipan envenomation. The vomiting had ceased and his blood pressure was now back to 150 over 75. He was fully conscious and had no trouble breathing. "Stable enough for medical ward," his accident and emergency report concluded. Clive seemed to have survived.

"I was feeling so much better," he said. We were retracing our steps along the river. "At about seven-thirty, we all decided Blanche and the girls should get back to Mareeba. We had an old friend coming for a barbecue that evening. I was feeling pretty fit and we didn't want to let him down."

At eight o'clock that evening, however, one of the nurses had trouble getting a blood sample from Clive's arm. When she withdrew the needle to try again, a thin dilute that looked nothing like blood dribbled from the puncture. When the hospital staff checked inside Clive's mouth, they noticed the same substance leaking from his gums. Blood was seeping from the membranous areas there, and from his nose and anus. When they turned Clive over, the bloodstains on the sheets were scarlet.

Clive had survived the neurotoxin's assault on his respiratory system, and his kidneys had endured the muscle damage. But the taipan venom contained another life-threatening component. Shortly after the bite, the venom had triggered the clotting cascade. This complex biochemical process is normally activated to create clots in a measured response to cuts or bruising. The venom, however, had triggered it on a massive scale, littering the bloodstream with rogue clots that threatened Clive with thrombosis. He had been spared the thrombosis and the clots had now dissolved; but the side-effect of the clotting frenzy had been to exhaust absolutely the blood's subsequent ability to clot. The relevant chemicals had all been used up. Clive's blood was now dangerously thin and the present danger was the absolute antithesis of thrombosis; that the capillaries might begin leaking blood into the body's internal cavities. Clive was now facing cerebral hemorrhage.

They raced him into intensive care. They upped his levels of antivenom but the priority was to restore his clotting ability. For hours, they IV'd him with fresh blood plasma. Initially, all they could do was keep pace with Clive's blood loss. He came to in a freezing fever, thinking he was about to die. But they wrapped him in thermal blankets and he didn't die. The bleeding began to slow at eleven o'clock that night.

They discharged him on the Wednesday evening. On the Thursday, the *Cairns Post* gave him the front page. The phone didn't stop ringing. An American television show, *Emergency 911*, even came to Mareeba to make a programme about the man who'd done nothing more routine than go for a swim in the river, and almost died on the bank. That was the true horror of Clive's experience; the way in which the taipan had subverted an activity as normal, as homely as a midday cool-off, corrupting the very landscape in the process.

We had returned to the foot of the path that had brought us down from Middlemiss Street.

"I've seen snakes since then, you know," Clive concluded. "Mostly the harmless water snakes but the odd taipan too, including a big one among the mango trees upstream, where the walking track runs to the old town swimming hole." Clive was conversational. "I sometimes wander up there if my knee's up to it. I guess it's about a mile. There's often snakes around. Just got to keep your eyes open, that's all. The way's wide enough. You're interested in snakes, you might want to take yourself up there."

I doubted I might. Walk along the banks of a river that had

already seen two deaths from taipan bite, and one extremely close call? Me? On my own and *away* from the hospital? At the time of year that taipans were known to be aggressive? I looked to where the track ran along the near bank of the river to disappear upstream in a flare of sunlight, and a shiver ran through me.

It was then that an elderly couple hove into view along the walking track. They were white-haired and walked in single file. They wore sunhats. She led the way, allowing her hands to drift through the overhanging foliage. He followed behind, tapping contented fingers upon his stomach. Occasionally, he would dawdle and raise his head towards birdsong. She, who knew him for a lingerer, felt she must jolly her companion along and often addressed him, but did not turn to do so. His responses were amiable monosyllables.

This frail couple had just emerged from the place I believed I could not tread, and they appeared enraptured by the experience. My nightmare was their stroll in the country, an afternoon enchantment even down to the yellow wildflower tucked behind the woman's right ear. Their appearance shamed me. I tried reminding myself that the walking track was said to be in good condition. That it was even concreted in sections, which was why the local kids so liked hiking it. That it wasn't as if I'd be wading knee-high through grass or anything. Keep my eyes open, tread carefully, carry a hefty stick. That I could even wear the heavy leather boots I kept in the car and think up some excuse—like I was wearing them in for what, a trekking holiday in Nepal—if anybody asked.

I'd love to, I told Clive. But I was busy, I said. And I followed him up the path to Middlemiss Street.

Later that afternoon, I remembered David Williams's snake collection. Back in Cairns, David had told me how the wildlife people had recently charged him with neglecting his snakes but had agreed that Brian Starkey, David's mate, should look after them in Ravenshoe, barely an hour south of Mareeba, until his court case came up. I was feeling pretty ashamed of myself when David's snakes came to my rescue. OK, so I didn't have the nerve for that river path, but at least I was up for seeing live taipans. Which was more than I'd done since I'd arrived in Australia. Brian Starkey answered his phone. Sure, he said. He'd meet me outside the old pet shop on Ravenshoe's main street.

As the road wound uphill, tropical crops gave way to forest and long expanses of windswept grazing land. At almost 3,000 feet, the small timber town of Ravenshoe was the highest in Queensland. There was little activity on the wide main street, just the flash of the white shorts and shirts of the lawn bowlers striding the greens in Memorial Park.

I drew up near the pet shop, a purple, wooden-fronted building tucked between the pharmacy and the Highland Bakery. A man with long wheatish hair, a straggly beard, a saw and hammer clamped under his armpit and a thicket of nails protuding from his missing teeth (the result, I was to discover, of a playful kickboxing encounter with a friend that had misfired) was leaning some pieces of plywood against the wall of the pet shop. Brian Starkey might have been one of the ugly men in a seventies rock band.

"Bastards," he said succinctly, peering at the pet shop window. A crack about three feet long ran diagonally across the window pane to a roughly circular hole about the size of a football. "People got nothing better to do than break windows," he muttered. "Sit around, get drunk, throw things." What he knew he couldn't allow, given the contents, was inquisitive types getting inside. Although he wouldn't half mind seeing their faces . . .

"You found us OK?" said Brian, pushing the door open. He came from down south. At fourteen, he had worked at the reptile park in Gosford, New South Wales, where all the aspiring herps did time. He had recently moved to Queensland, and had been working for David Williams, mostly collecting snakes in Papua New Guinea.

I followed Brian into the gloom of the pet shop. A shadowy creature scuttled into a corner. I must have jumped for Brian was quick to reassure me.

"That's just Lurch," he said, reaching for the lights. Lurch was a three-foot goanna lizard that enjoyed open confinement there. "Also David's got a couple of small estuarine crocodiles in here," Brian added. I remembered what estuarine crocodiles had done to the woman at Daintree, and I jumped again. But Brian really did mean small. In the flickering strip lights, I saw that the crocs were about two feet long, and lay disconsolately in a pool of sorts that had been arranged for them in a corner of the pet shop.

In the light, I could also see that the Ravenshoe pet shop had had its time. Sheets of damp-stained plasterboard bulged dramatically from the ceiling, giving an unintentional draped effect to the room. Old wooden cupboards, handsome, decorative things with corniced edges and panelled doors, ran along the walls and were also arranged to form a central island. Piles of old newspapers lay on the bare wooden floorboards. The room had a thick, musty smell, the smell of confinement and of reptile shit. Glass-fronted wooden boxes had been stacked on the unit tops.

"Look around while I sort the window," said Brian. "You'll find them all there."

The boxes weren't labelled. They looked a bit like the childhood boxes we used to keep guinea-pigs in. So did the floors of the boxes, lined with old newspapers in the same homey way. But what slid across the appeals for Tablelands women to undergo breast screenings were taipans, and it was king browns that concealed the local Aussie Rules match reports, and tigers and copperheads the ads for discounted panel vans. There were even rare inland taipans, whose venom was reckoned to be the most poisonous of all land snakes, rubbing their bellies against announcements for meetings of the Mareeba Rotary Club.

Many of the snakes had awoken to the vibrations from Brian Starkey's hammer blows. They were alert, prowling their boxes, throwing strange vermicular coils and pressing their white bellies against the glass. I stretched out my hand to touch through the pane the shingle coils of a king brown, or the girasol eyes of a rust-coloured taipan through the pane.

The same strange FNQ subversion again. Just as seagulls went for you, magpies dive-bombed you and a swim in the river turned into a near-death encounter in this part of the world, so the Ravenshoe pet shop had gone X-rated. It had thrown out the cozy dog leads and the chews, the chrome cages and the cat baskets and the worming tablets, the hamsters and the goldfish, the bags of dog meal and the Bonios, the gerbils, the cockatoos and the guinea-pigs, and hosted instead what must have been, snake for snake, among the most venomous collection anywhere in the world.

"You done?" asked Brian Starkey as he secured the window with a final piece of plywood. I looked at a taipan one last time and stepped out into the sunshine. One of the lawn bowlers, bored perhaps by his part in the game, looked up and saw us. For a moment, he may even have wondered just what was going on at the old pet shop.

18. AFRICA V

The Ikoma driver, one of the rare survivors of a black mamba bite . . .

—Peter Matthiessen, *The Tree Where Man was Born* (1972)

Normally, the mamba's bite, the most deadly known to man, kills its victim in just five minutes.

—*Newsweek*, May 9, 1977

The zoo keeps a supply of antivenins but as some snakes, such as the black mamba, can kill in under an hour, there would be little time for treatment.

—*Official London Zoo Guide* (1983)

The mamba did not acquire worldwide notoriety overnight. A Zulu term, mamba had been in common usage in South Africa for much of the nineteenth century, but it was slower to acquire international recognition. The word had barely reached Kenya by the 1890s—William Fitzgerald wrapped it in inverted commas to denote its unfamiliarity when he wrote of his encounter with a green mamba in a tent by the Galana River on Christmas Day 1891. Of the black mamba, Fitzgerald made no mention whatsoever, preferring to describe *Dendroaspis polylepis* as the tree cobra. And as late as 1911, the mamba did not merit the least mention in the *Encyclopaedia Britannica*.

When the mamba finally came to prominence beyond its range, it was *Dendroaspis angusticeps*, the smaller, more retiring and less venomous green, that made the early running. During the 1920s and the early 1930s, London Zoo's official *Guide* referred only to the green mamba. "The most dreaded of the

African snakes," it read, "as it is a strictly arboreal species and frequently bites natives in the shoulder as they push their way through the forest paths..."

It appears that the zoo acquired its first black mambas in 1935. Nineteen thirty-six was the year in which its *Guide* first made mention of them, and immediately gave them equal top billing alongside the greens: "The mambas, black and green, are perhaps the most deadly of all poisonous snakes; they inhabit Africa," it read. The black mamba was making up ground, but it was not until the 1950s that it could plausibly claim exclusive rights to those prized ophidian titles, tributes rather than slurs—the most deadly, the most notorious. It was in the 1950s that C. J. P. Ionides began sending snakes, black mambas among them, to London Zoo. He sent one in 1952 and another the following year. In the course of 1954, he donated some forty snakes, including green mambas, puff adders, boomslangs and a gaboon viper. But it was the arrival of that year's black mamba which caused the greatest stir. For the word was that Ionides had sent the zoo a mass killer.

Ionides had caught her on March 7, 1954, in the Mtwara district of Tanganyika. She was, by his own account, a magnificent specimen, over ten feet long and, "a beautiful dark purple above and white below with a lovely bloom on her." He had found her curled on a low branch above a path, having been led there by a local headman who explained that the exact spot had seen as many as nine fatal cases of snakebite over the last two years. Ionides's black mamba was held responsible. Even today, the dark deeds of Ionides's snake are commemorated on the black mamba pen at London Zoo's reptile house—"One that came to the zoo in 1954," it reads, "was supposed to have killed six people"—even if the zoo's headcount is rather more conservative than that of Ionides's headman.

By the early 1970s, the black was easing the green out of the shared top spot. "Among the most dangerous," the Zoo *Guide* now read, "are bites from the black mambas." The black mamba had become a byword for serpentine hostility, aggression, speed and virulence. It was a measure of its awesome reputation that it was the undisputed snake of choice to star in two novels of the 1970s, a decade that spawned the singular *Jaws*-type fiction genre to which they belonged, thrillers obsessed by the notion of wildlife's insatiable appetite for mayhem, and its ability to deliver it.

Alan Scholefield's *Venom* and John Godey's *The Snake* have a black mamba loose in a London house and New York's Central Park respectively. One, destined for the London Institute of Toxicology, is mistakenly delivered to a young boy expecting to add an innocuous house snake to his pet collection; the other, won "from some Greek in a poker game," is smuggled through New York's docks by a returning sailor. While Scholefield's mamba kills three, the body count in the Godey book is four plus four wounded, which could be said to be New York for you. Both books have characters saying measured things like "I'm inclined to think it's the most dangerous snake in the world." And both *The Snake* and the movie version of *Venom* conclude with the genre's statutory cliché, the monster safely dispatched but not before she had laid an unseen hatch of eggs in some safe, warm spot.

Venom was published in 1977, which happened to be the same year that two men took serious bites from black mambas in Africa. One, which occurred near Pretoria in March, made the international press. The other took place six weeks later at Kilifi, in Kenya.

JUNE 1996

Francis Ngombo had the agape eyes of some small night mammal. *You what?* they seemed to say, countering an otherwise *I know, I know* appearance. Surprise and serenity combined, Francis made an unsettling impression. He was a slight man, with cropped frizzy hair, unevenly greying. His skin was almost translucent; it stretched over a fleshless skull that seemed delicate as eggshell. His thin lips and chiselled nose were worked, it seemed, from teak. He had the face of the coast. It spoke of Arabia and clove-scented lateen dhows. He arrived at first light, looking like a butterfly collector or an angler; he carried a six-foot bamboo from which a small noose of plaited brown sisal hung, and a few grey sacks, folded neat as pillow cases, draped over his laundry-girl shoulder.

I had been waiting for Francis at the Gedi junction, where the Watamu road meets the main Mombasa highway and where a brand-new gas station stood. It doubled as a shop, a most un-African shop, with gleaming aisles and neat stacks of imported processed groceries and, the station's pride and joy, a shiny chrome steamer in which a litter of submerged hot dogs lay, jostling for space like river lumber. I had left Barbara's long before breakfast, so I fished two sausages into waiting rolls and scored them from top to toe with a rib of mustard that looked alarmingly bright in the grey morning.

"Enjoy your dogs." The man at the counter smirked as he spoke, doubting the sense of the line I suspected he had found in the steamer's instruction manual—under "customer relations"—but relishing the sound of it anyhow.

We crowded onto the first southbound matatu in a rush of schoolchildren and young mothers. Everybody seemed to understand the significance of the noosed stick and the sacks; even in the crush, the snake man and the mzungu managed to gain themselves a little space. Francis had been interested in snakes, he told me, since his childhood. Francis would linger long after the other children had fled the snakes they came across in the village, transfixed by their markings and movement until an adult came by to wrench him to safety. The distinctive sound of an unseen snake on the move, an assured, fluid rustle, soon captivated Francis. Hearing it would cause him to stop and watch intently. With his keen eyes, he learned to locate such snakes, glimpsing brief sections of their polished bodies as they poured themselves deeper into the bush. He soon learned which he must avoid, and which he could safely catch. By the time he turned twenty, he began to ignore many of the superstitions about snakes which even then cowed his parents, and he resolved to become a snake man. He learned to catch dangerous snakes which he sold for a few shillings to the snake parks that were opening, along with the tourist trade, all along the coast. He hung around the white snake men, learning what he could from them and picking up occasional work.

His father was incensed; working with snakes was no profession for his son.

"At the beginning, he would not eat with me," said Francis. "He thought it evil to have such an interest. He thought I was becoming a mchowi. I wasn't welcome at home."

In time, his father's objections began to abate. Especially when the right snakes brought the hard-pressed family a few welcome shillings.

Francis, who was now forty, had seven children to feed and hunted at every opportunity. We left the bus at Masngoni: "the walking place." A few roadside mud and makuti thatch huts huddled under coconut palms where wood fires spiralled purple smoke. We walked through the village, flushing out desultory accounts of snake sightings from the few villagers who were about. Yes, there was a puff adder around yesterday, a man told Francis. The man wore a football shirt emblazoned with the word "Carlsberg," and poked optimistically at the white ash of an old fire. "But somebody killed it already. The dogs took the body. Someone else the head. And one of the children saw a green snake, a mamba perhaps, in the cashew tree over by the school a couple of days ago."

So Francis strode out of the village with a dawn jauntiness, explaining how there was no better time of day for finding snakes.

"Before it gets too hot," he explained. To scan the high wall of dense green bush, thorn and overhanging trees that flanked one side of the dirt track, Francis walked sideways, crossing his legs in music-hall style. The tightly packed group of inquisitive children that trailed in our wake mimicked him. After a few minutes, Francis stopped abruptly. He lay down his sacks and lifted a cautionary finger. I repeated the gesture to the children, who repeated it in turn to the younger among them so they were thicketed in raised fingers. The only sound was of lorries on the distant highway and a wind off the ocean tugging at the corners of Francis's sacks. Slight Francis appeared to float towards the bush, slipped two fingers inside the sisal noose to enlarge it and then calmly moved the stick towards the bush.

I did not see the snake until the noose was suspended in front of it, framing its head in a locket oval. It was draped along a branch and was about two feet long. It was mottled grey and brown, and it did not move.

"Link-marked snake," whispered Francis. "Not very venomous."

"Lick mark snake," the children told each other excitedly. Francis calmly slipped the noose over the snake's head, and hauled it clear of the bushes. It thrashed with a metronomic rhythm, throwing a shapely "s" from side to side with something of a windshield wiper's regularity. With one hand, Francis grasped the snake tight behind the head, causing it to wrap two defensive bracelets around his wrist. With some deft fingerwork, he loosened the noose, freed his wrist of the snake, opened a bag and dropped the snake in. The children stood transfixed, so much so that they had forgotten to lower their cautionary fingers. Only as Francis pulled the neck of the bag tight did they begin, one by one, to do so.

Within minutes, Francis had bagged another link-marked snake. Seconds later, a hissing sand snake got away. If you knew where to look, there were evidently plenty of snakes about. In a six-month period in the 1970s, Malindi snake man Mark Easterbrook and his team had collected over 5,000 black mambas—an average of around thirty a day—in Kenya's Kerio Valley. More recently, he had collected tens of thousands of green mambas along this very stretch of coast. The coast was alive with greens. In a celebrated incident to the south in 1952, one had even found its way into the cabin of a passenger aircraft awaiting take-off from the airport at Dar Es Salaam in what is now Tanzania. It had draped itself around the pilot's neck, and struck at his tunic as it fell to wrap itself around the control column. Take-off was postponed.

"Mamba!" yelled Francis suddenly, diving into a nearby mango tree. My heart shuddered as I glimpsed a green snake pouring through the branches. "Mamba!" yelled the kids but they were querying the sense of their exclamation even as they uttered it. For mamba happened to mean crocodile in Swahili, and, no question, there was no crocodile in the mango tree. They began to laugh. The snake, meanwhile, moved very fast but erred by making regular stops to assess its predicament, which was when Francis noosed it and brought it to the ground.

"Ah," said Francis. "Boomslang, not mamba." The snake was a lovely iridescent green, about two feet long. Francis handled it with great care. Unlike the mamba, the boomslang's fangs are set far back in the jaw, making an effective bite unlikely, but its potent venom does not make for complacency.

"*Mgangarudi*," said Francis to the children, giving the snake its Swahili name, *Doctor, go home*. The morbid inference, that there was nothing the doctor could do about such a bite, silenced the children.

The snake's Swahili name was perhaps melodramatic, but it was considerably more informative than the Afrikaans; boomslang means "tree snake," and Francis's every catch that morning had come from the trees. With the significant exception of puff adders, herpetological action was scarce at ground level. Francis had not so much as glanced at his feet all morning.

I remembered how Fitzgerald had called the partially arboreal black mamba the tree cobra, a name which seemed to recognize the snake's true element. The black mamba excelled in
the trees. Eugene Marais, the South African naturalist, described the mamba moving through branches as "an arrow from
a bow." "No wingless creature," he marvelled, "more nearly ap-
proaches flight."

Trees may have been the mamba's stage; they were also, however, its place of ambush. *The London Zoo Guide* had claimed
how greens struck from the trees, biting the shoulders of passing Africans. And Ionides's 1954 serial-killer was supposed to
have operated from a branch above a well-trodden path.

The same attack strategy had long been attributed to the
great mythic serpent of Central and Southern African lore, variously named as Mukunghambura in Kenya, Muhlambela in
the Transvaal, Indlondlo in Natal and Mbobo in Zimbabwe.
This vengeful monster of the trees was to Africa what the vampire once was to Central Europe, the wolf to Siberia and the
bogeyman to the American South; the bane of children's nightmares. In its most extreme form, it was hundreds of feet long,
had rainbow markings, a crest of feathers and a bleat like a distressed deer to attract its victims. But it was most feared for its
habit of launching fatal blows at the back of its victims' heads
from its place among the branches.

Nor, apparently, was the great serpent a fiction. There were
educated men who believed it actually existed, if on a rather
more credible scale. Livingstone had heard talk of the monster
snake on the Lower Zambesi, where it was reportedly known as
Bubu and grew up to twenty feet in length. The male crowed
like a cock. The female clucked. And it had a fleshy red crest.

A Dr. Shircore of Nyasaland even claimed to have a bony
portion of one such crest in his possession. Shircore wrote in
the *Journal of the Royal African Society* in 1944 of this monster
snake's habits. "It moves with great rapidity," he claimed, "and

it is tree-loving like the mamba, which it resembles in several respects." He continued,

> Periodically, it selects a haunt near a road, or path, which traverses some forest area, and, hidden in the dense foliage...coils its tail around an overhanging branch. From this it strikes at the head of any man, or animal passing below. When this happens, travellers are warned off the road, or, if they must go through, observe the greatest caution. The old African custom was to carry a pot of hot porridge or water on the head, so that the snake in striking might be burnt and desist from attack, or kill itself...so great is the dread of this snake that African Telegraph Linesmen, whose duties may bring them into the danger zone, take every precaution to enter, and despatch their work, silently, and get away as quickly as possible.

The day grew hot, and began rendering me impressionable. I noticed with the first stirrings of alarm that there were plenty of trees about. I also noticed that Francis was positively making for the trees. He had begun to weave his way forwards, not from exhaustion but so as to stepping-stone from the shade of one tree to another. Fearful of what the mangoes, cashews and coconut palms might harbour, I followed with reluctance. Occasional mud and thatch huts stood among the trees. Against one such hut, a man had propped his bicycle. He wore a white Muslim skull-cap and a long *jellaba* gown, and was sitting cross-legged upon the ground. Opposite him, a woman was listening intently to the sounds emanating from the gourd she was shaking.

"A mchowi," whispered Francis, clearly rattled. "He has come for advice from her. The noises the gourd makes tell her the secrets of the man's difficulties."

She looked up at our passing. Her eyes seemed far away, but they caused Francis to quicken his pace. "She will know exactly what we have been speaking about," he whispered. "When we go back, we must take a different path."

Francis knew some of the villagers in these parts, but mchowis tended to take exception to strangers who caught snakes, the strangers whose white friends claimed there was nothing to snake-catching, nothing but a steady hand, and so mocked mchowi magic.

We drifted into a village Francis knew, a place of dark, tramped earth surrounded by a few huts, a couple of tethered cows and some chickens. Five men had arranged themselves on old seats and tree trunks in a rough circle, and were upending old Johnson's Baby Lotion containers and Gordon's Gin bottles. They were drinking fermented coconut palm juice called *mnazi*, the "toddy" favoured by Fitzgerald's snake man. Children idly rolled calabash gourds across the earth. Bare-breasted women propped an elbow on a knee, a chin on a fist, and stared into the trees.

The mnazi was sour, but it was Francis's lunch. It fortified him. He became expansive and garrulous, and regaled the villagers with stories of run-ins with giant mambas, thrilling escapes, holes in sacks. The villagers listened to his every word. Not many visitors came through here, certainly not many snake men, and they had abandoned any pretense at being busy years ago.

"They say a tsasapala, a black mamba, came through here only a few days ago," Francis told me. "One of the women saw it. It went straight between two babies sitting on the ground. Fast like a matatu. The babies didn't see it."

Francis smiled and drained his mnazi container. The drink had turned his eyes red, and the usual look of surprise had given way to a new recklessness.

"So," said Francis, rising to his feet with an uncharacteristic swagger. "Let's find that black mamba I promised you." He handed over a few coins to one of the men, and pushed off into the forest with a flick of his wrist.

Strange, I had hardly touched the mnazi but a light-headedness, an apprehension, had settled upon me. Francis was walking ahead of me now. He advanced in stops and starts. The sun was very hot. It sat upon my head, drawing down a silence. Even the insects seemed to have fallen quiet. Francis was walking down a path that fringed plots of maize, cassava and sisal. Where the path met a wall of thick bush ahead, it turned right. But Francis kept straight on.

"They never cleared this bush," he said. "To the villagers, it's a sacred place. Also a good place for mambas." He pushed his way between two bushes and disappeared from view. I followed, feeling more lightheaded than ever. I pushed aside the bushes to reveal a place of deep shade. Cobwebs were spread like traps across the panes of space amidst the foliage. Black branches ran like cracks across my vision. There was thick leaf mould at my feet. Suddenly, something moved in the undergrowth and a white-fronted bird flew up at my feet. I jumped.

"Coucal," whispered Francis. "White-browed coucal." I remembered the connection between this bird and the black mamba, and I jumped again.

That was when it began to get darker, as if the sun itself was being blocked out, halting the thin filter of light through the foliage. The bush was thickening about me. Things were growing much faster than they should, bristling with new buds and unfurling leaves. Suddenly, I could sense Francis stiffen ahead of me. And he lifted that same cautionary finger. Then he lifted another, and another, until his every digit was extended like a showman telling a joke. Francis had left his snake stick outside the grove. So he began picking the snakes off nearby branches with his bare hands. The snakes were lined up like rolled belts in a clothes shop, which made it easy for him. He just dropped them one by one into a bag. When one bag was full, he slung it over his shoulder and reached for another. Some of the snakes he rejected, replacing them carefully on a branch and turning to whisper the name of some dead relative he had just discovered the snake to be. I didn't know much about the snakes he was collecting, but Francis's work impressed me.

Francis had left the black mamba till last. It was much higher in the tree. It had a lovely red crest, far more splendid than those of the scraggy chickens in the village, and a sinuous body that went on forever. It was wrapped round the trees in a rising spiral, like the approach road to a multi-story car park. It clucked a lot, and was trying to strike at the back of Francis's neck but Francis knew his snakes, and he kept turning to face it. The snake kept adjusting his own position, but Francis just smiled and matched it move for move. I then noticed that Francis was stirring a pot of hot porridge with his free hand. I was so busy watching him that I suppose I must have taken my eye off the mamba. I only realized where it must be when Francis turned and turned until he was looking directly above my head. His reckless eyes immediately turned to surprise and then to fear. Then I felt a blow fall on the back of my neck. And I screamed.

Francis was kneeling over me, all the serenity gone.

"Are you all right?" he asked. I had taken a few thorn scratches tumbling out of the grove. I lay on the path, trying to clear my head.

"Something got me in the neck," I told him. But even as I said it, I realized I'd merely been nipped by some insect. It was not a black mamba which had poisoned me. It was the Lariam.

FEBRUARY 1997

Pete was speaking into an empty hotel lounge. He remembered thinking, through the dizziness, how his Sunday snake-handling superstition was a strange thing in a rational man when Jan ran up the Takaungu steps with David Corroyer. Pete wasn't given to unreasoned courses of action. He trusted in the application of good sense. The empty syringe and the few glass vials lying

beside him testified to that. So did the words he greeted Jan
and David with; no hysterics, no panic, just the fact that he had
injected the little antivenom he had to hand, and that he feared
that it was not going to be enough. He was sitting on the sofa,
hands planted either side of him to keep himself upright. His
face was grey. He had recently thrown up. Corroyer brushed
the vomit from Pete's chin, picked him up and laid him on the
back seat of the car. They left the house in a pall of dust as
Corroyer raced down the dirt track that led to the highway. The
tires threw up small stones that hammered at the bodywork as
they passed through fields of sisal and groves of cashew trees.
When they reached the highway, Corroyer turned on his lights
and hammered the horn every few seconds. From the roadside,
it sounded like an unpracticed wind section passing in a hurry.
The car shot through villages, weaved between matatus and put
up scraggy squadrons of roadside chickens. Jan leaned over the
back seat, where Pete was drifting out of consciousness.

Now that he was in mortal danger, Pete's life might have
been expected to flash poetic before his eyes, his grandfather's
arrival in Africa, footing it to Nairobi before the train, the
earthquake fault yawning open to swallow his crib, the trips to
catch cobras with his uncle and the puff adder hanging from
his uncle's arm on a moonlit New Year's Eve, his first elephant,
his first thoughts of a snake park, the ferry ride to the hospital
after his first bite, the propeller screws boiling the water in
Kilifi creek; and now this, his Sunday snake. But Pete was not
given to such coloured reminiscences, not even now. Pete was
not that kind of man.

Pete could tell you all about keeping snakes in captivity. He
could show you how to get captive snakes feeding and how to
keep the mites at bay. He could explain his own design for
snake cages, and how to treat the wood for white ant. When it
came to the practical stuff, there was nobody better than Pete.

But imagination wasn't Pete's forte; he'd say so much himself. His snakes were his work, his income. He was a snake

professional. At his most enthusiastic, he regarded them as fascinating reptiles. What he wasn't was a snake visionary. When it came to snakes, he'd never allowed his thoughts to run away. Kept his imagination reined in. That way, he remained immune to the malevolent beauty of snakes, an inherent suggestiveness that could do your mind in, invade your dreams. Snakes could break a man. And Pete dealt with snakes every day. He'd handled thousands of venomous snakes in the course of his life, and he'd only managed to do so without thinking too much about it. You couldn't afford to think too much about handling snakes. You couldn't let the imagination go. Otherwise, they'd drive you insane, as snakes were wont to do. So Pete just lay there as the road pummelled him through the suspension and the African sun flickered against his eyelids, and he told himself how stupid he had been.

They reached the hospital in record time. A few members of staff were enjoying an uneventful Sunday morning when Corroyer ran in hollering for a stretcher. By the time they got Pete inside, the anesthetist was on hand with the antivenom. Soon, however, Pete had exhausted the hospital's antivenom supplies. By mid-afternoon, as word spread that the Kilifi snake man had taken a black mamba bite, a steady stream of antivenom vials were delivered to the hospital. Some were out of date and only being kept in case the dog got bit, but it all went straight into Pete Bramwell. His system needed every drop it could get.

As the sun went down, Pete's condition remained critical. They were having trouble keeping him breathing. His lungs collapsed at ten o'clock that night. There was talk of putting him on the iron lung that a local diving concern kept in case of underwater accidents, but the owners were up on the Northern Frontier and had taken the only key. So they put him on a ventilator which succeeded in getting him breathing again. When his lungs collapsed for a second time, at two o'clock in the morning, the anesthetist thought the snake man was about to die. But again they managed to get him going. By dawn, there were signs his condition was steadying. The donated vials of antivenom were now being stacked by Pete's bed rather than going directly into his system. He opened his eyes at nine o'clock on the Monday morning, twenty-four hours after he had been bitten. He was discharged on the Tuesday evening.

He was now sitting in a cloud of smoke, behind a low barricade of coffee cups, ash trays and cashew nut wrappers. In his old coat and pebble glasses he did not look like the survivor of a black mamba bite. He was an old man in a cold country that was not his own as the best story of his life came to an end.

The only thing was it was not the same story I'd heard in Kenya, the story I'd pieced together from snake-park attendants, hotel receptionists, locals and barmen. Not the same story at all.

JUNE 1996

I walked back to the road behind Francis, feeling foolish. We had decided we had had enough of hunting mambas. On the matatu home, Francis fell silent. He seemed preoccupied.

"I am disturbed by people," he said when I asked him if anything was wrong, as if my recent strange behaviour had prompted some kind of intimacy between us.

"I shall show you what I mean," he added, pointing down the road into the near future.

For eight years, Francis had run his own snake park outside the ruins at Gedi, not far from the road junction. Until recently, business had been good. Visitors to the ruined Arab city, the atmospheric old palaces and mosques surrendering to baobabs and strangler figs, often dropped in for a look around.

"Three months ago, I gave the monthly rent—500 shillings— to the owner's son," said Francis. "He was supposed to give it to his father. He now denies I ever gave it to him in the first place. I will make the debt up when I can, but I will never hand over money to his family without a receipt."

The park, perhaps half an acre of flat earth, was ringed by a low barbed-wire fence. Its fancy wicker and makuti thatch entrance arch, which Francis had constructed, had recently come crashing down. We stepped over it and into the snake park. There were no buildings, but the forest canopy provided ample shade. Francis had worked assiduously at the park's gardens. On every side were neat beds ringed by stones where flamboyant grew. He had built a pond for his turtles. Among the trees were turned-over cane tables. On these, Francis indicated, the snake cages had been placed. He pointed at the cages, with their glass or gauze fronts, now piled haphazardly in a corner of the park. There were tears in his eyes. For a single month's rent, his business had been wrecked. And until he paid, his landlord was going to keep it that way.

"I am disturbed by people," repeated Francis.

On my way back to Plot 28, I stopped off for a drink at Ocean Sports. The bar was empty. The barman was staring distractedly at the surf, a glass wrapped in a dishcloth.

"Been fishing?" he asked, anxious for company.

"Hunting mambas," I replied.

"Now, that's not one of the lines I've always wanted to say. Mambas? You mad or something? You know there was a chap got bitten down at Kilifi."

"Yes, I heard some—"

"Terrible, it was...Get this, he's lying there, paralyzed, and he hears the doctor telling his wife that he hasn't got a prayer, that he's going to be brain damaged and they may as well pull the plug. And he's thinking, *Christ! Don't do that to me.* But there's nothing he can do to show them he's aware. He can't talk. Can't move. Can't do one simple thing. And they're about to turn him off. So he knows he's got to show them he's alive, somehow. And soon. So he does all he can to move something. A finger, a toe, anything. But he can't do it. That evening, they bring a priest in to give him the last rites. They're going to turn off the machine, see. And all his family are there standing round the bed, and they're about to kill him. And he's thinking, *this is it.* And the priest blesses him. And as his wife leans down to kiss him goodbye, she sees a single tear spring from his eyelid. And that's when they realize he's alive. And the doctor takes his finger off the life-support switch. And a couple of weeks later, he's out of hospital. You know what he does? He takes his shotgun, goes down to his mamba pen and blows the lot of them to pieces."

"And you don't know the man's name, do you?"

"Sure I do," replied the barman. "Bramwell. Pete Bramwell. Moved to Scotland some years back. Let me call up his son-in-law. See if he can give you an address."

I asked for another beer and offered the barman one while I was about it.

19. INDIA III

At Sangli, a town in Maharashtra, a man was kneading six-inch statuettes of hooded cobras from a bucket of wet, mocha-coloured clay. Eyes and mouth he gouged with the sharp point of a scissor blade, finishing them off with little forked tongues cut from red felt, and standing them on his stall to dry in the first, flat rays of the dawn sun. It was a day for making snakes.

Autorickshaws disappeared into the clots of thick dawn mist that had formed along the street outside Sangli's bus station, and emerged dangerously askew, the drivers rubbing their eyes. I had left Madras at noon the previous day on a brown, clanking train that rolled north-west across India for sixteen hours. At the rain-smeared windows, occasional sacred bullocks, their horns painted blue and gold, loomed large and sudden, and seemed quite as surprised as the passengers at their unexpected convergence in the depths of Andhra Pradesh. In the compartment, a young man read about pre-shipment credit in *Borrowing from Financial and Banking Institutions* while an elderly man cast mournful glances under the cloth which shrouded a domed birdcage.

"A green parrot," he told me. "My wife's. She died only a month ago. The doctor told her to stop eating strong curries. Nonsense, she informs him, and then she dies clean away. After a strong curry." He dabbed an eye with the corner of his kerchief. The parrot was silent.

I had left the train in the early hours and continued my journey by a series of decrepit red buses to Sangli, where the clay cobras with red forked tongues reminded me that it was the eve

of Nag Panchami. India's cobra-worshipping festival is celebrated on the fifth day after the full moon in the Hindu month
of Shravan; July or August in the Roman calendar. Snakes are
common at this time of year, and the monsoon is firmly established. A good crop looks likely. As if to celebrate the apparent
connection between snakes and the promise of plenty, cobras
graven from clay, wood or dough are worshipped and their image daubed in turmeric, wet rice powder or cow dung upon the
doors of houses in many parts of India, but particularly so in
the shadow of the Western Ghats, in the states of Karnataka
and Maharashtra.

In the 1800s, schools in Maharashtra towns like Sangli,
Satara and Kohlapur closed on Nag Panchami. Pupils dressed
up and joined a grand procession led by richly caparisoned elephants and horses, all to the beat of kettle drums. Even elegant
Maharanis, accompanied by pipers, set out to worship at the
termite hills which cobras had made their homes, scattering the
petals of oleander and lotus flower, water lilies and jasmine at
their entrances.

But Sangli was not the end of my journey. A final red bus or
two would bring me to the town of Battis-Shirala, which translated, unaccountably enough, as the thirty-second place called
Shirala and which had been renowned for its Nag Panchami
celebrations for centuries. Legend claimed a local god once be-
rated the townsfolk for worshipping mere images of the cobra,
since when they had preferred to work with the real thing.

As our red bus approached Shirala, I could see young men
scattered among the fields beneath a sky puffy and bruised with
rain-laden clouds. Some squatted bare-chested, their *lunghis*
rucked round their waists while the heads of others showed
among the stands of maize and sugar cane. In most respects, it
was a typically Indian landscape but the various figures in it
shared a sentinel intensity that seemed at odds with the mundane routines of crop-tending.

The bus rounded a bend, passed a riverbank where women

were dashing wet, shining clothes against a rock, and pulled up beside a vine-clad banyan tree in the middle of town where purple men thronged. They were caked in *gulal* dye, the happy powder that is the confetti of Indian festivals. Some held stout branches before them from which brown hessian bags hung heavily. As they walked, they were chanting something triumphant in Marathi.

"Whose cobra? Our cobra!" one of them translated helpfully as he stopped to introduce himself. Sayga Ghatge was fifteen, and the only English speaker among his group. The Ghatges and their friends, Sayga explained, had caught their cobra just that morning. Which explained the vigilant men I had glimpsed in the fields.

"We see you later," said Sayga, running to join his group which was fast disappearing up the street. I waved and, dodging handfuls of high-spirited gulal, went straight to the Kismaath Hotel. The man looked confused when I asked for a room.

"We're a hotel," he said patiently. "We serve food." So I tried the Shirala Guest House, which turned out to be a bakery. It was like the Madras wall hoarding I had seen—"Expensive jeans; only 399 rupees"—which promised one thing and delivered another. Clearly, many introduced English words had not survived the journey to their new Indian home. Somewhere along the way—it might have been on the packet through Suez and across the Arabian Sea to Bombay, on the slow train to Poona, in the markets at Kolhapur or on the forerunners of that old red bus that would through the hills—they had strained, sheared and finally broken free from their native meanings.

"Ah, lodgings! Lodgings you are meaning!" exclaimed the young postman who stopped to assist me and introduced himself as C.Y.W. "Suggest you are trying Tirupati Auto Parts."

Tirupati Auto Parts was on the ground floor of an incongruously modern concrete block several stories high that seemed to have recently crash-landed on Shirala's outskirts. A round man emerged from the tangle of spark plugs, jubilee clips, fan belts and air filters in which he appeared to have been nesting to scratch his chin at my request before leading me up the stairs, the loose flesh on his flanks shuddering beneath his shirt. At the end of a long corridor, he turned a door handle and ushered me into a room that was empty but for drying onions and garlics. They smelt sweet and faintly intimate, and lay in such numbers that only occasional glimpses of the ridged and stippled concrete floor showed through. Garlic, I cheered myself; an excellent snake repellent. Still, I must needs know from Mr. Tirupati where I was expected to sleep, but he was already silencing me with a theatrical gesture that ran a great broom across the floor, sweeping a wave of compliant onions and garlics before it. He could also include a sleeping mat and a blanket in the price, he assured me, but insisted on payment in advance.

"And my name is Patil," he corrected me. "Tirupati is our famous pilgrimage site in Andhra Pradesh."

I stepped out into the corridor. It was open to the side of the building, affording views that were informative if uninspiring. I was on the edge of town, a motley huddle of buildings, smoke-blackened walls, tiled roofs, teetering chimney pots and tangled wires which stretched eastwards while a low-caste encampment of makeshift tents clustered around my building. Children played by smoke-blackened tents and the men chipped away at slabs of stone to produce rough-surfaced, shallow spice mortars. Black pigs, their pendant udders caked with mud, rootled in the rubbish. To the west lay groves of glistening trees and gentle hills and, rising among them, a Ferris wheel taking shape and the pointed tower of an ivory-coloured temple planted with orange prayer flags.

I left Mr. Patil to rid my lodgings of onions and went out to explore Shirala. But I had got no further than the bottom of the stairs when C.Y.W. cycled industriously past.

"Special deliveries," he shouted cheerily, then skidded to a stop, throwing his back wheel in a circle, to tell me I was going

the wrong way. "Cobra competition," he said, urgently thumbing over his shoulder. "Eleven hundred hours. Ubale Lane.
Most big attraction."

An animated crowd had gathered at the Ubale Lane competition ground, a fenced area of cropped pasture where six rect-
angles about the size of parking bays had been chalked out with
rice powder. There was a precarious wooden platform, covered
with a roof drape lashed to uprights, where townsfolk officials
held hefty ledgers and crowded around a microphone that
stood on a desk but did not seem to function. From the sanctuary of their sunglasses, and the ill-fitting suits into which they
were crammed, they effortlessly deflected competitor enquiries.

The competitors sat on the grass inside the enclosure. Beside
them lay sticks and round clay pots, head-sized. Each pot's
stoppered lid was wrapped with a brightly coloured kerchief, often orange or pink and embroidered or patterned with beads
and sequins. The lids were secured by a cord wound around the
lip of the pot in a manner that brought to mind greaseproof paper, elastic bands and jars of homemade preserve. Only, the
contents were more exotic.

The first heat began upon a whistle. To roars of encouragement, six pairs of competitors stepped into their allotted rect-
angles and undid the cords on their pots. Upon a second
whistle, they upended their pots into the rectangle. From each
pot, an enraged cobra tumbled to find itself facing a man goad-
ing it with a pot, the pot from which it had just been ejected,
while the other man restrained it by the tail. Livid, the six cobras flung instant hoods. As the men shook the pots just above
the snakes' heads, the cobras strained for height, and the roar
of the crowd rose with them. The suits stepped up to plant six-
foot measuring sticks beside the cobras before writing on their
clipboards what height each snake had achieved. Twenty-one
inches was good; twenty-three and a half inches would go on to
win the competition.

I confess I had not expected a how-high-can-your-cobra-go

competition. I suppose I had anticipated a rather more mystical atmosphere, a more spiritual experience altogether. The competition instead invited comparison with the Georgia rattlesnake round-up, the essential difference being that weight and numbers of snakes collected had won the prizes there.

I was being slow, of course; it had been a long night. I remembered the snake stones I had seen, the temple lingams and the words of that diligent nineteenth-century observer of Indian life, Rivett-Carnac, "the attitude of the cobra when *excited* (my italics) and the expansion of the head will suggest the reason for this snake representing...the phallus." All snakes were phallic but, as Rivett-Carnac implied, the hooded cobra was singularly so. Even, it struck me, down to the cobra's hood marking, commonly referred to as spectacles but more convincingly genital to my mind, complete with a handsome brace of testicles.

The cobra competition was a reminder that Nag Panchami was a fertility rite in the procreative sense as well as the agricultural one. "More than the usual license was indulged in," wrote Rivett-Carnac of the Nag Panchami festivals he witnessed at Nagpur, South India, in the nineteenth century. He wrote of "rough pictures of snakes, in all sorts of shapes and positions," being "sold and distributed, something after the manner of Valentines." These pictures he described as "hardly fit to be reproduced...the positions of the women with the snakes were of the most indecent description and left no doubt that, so far as the idea in these sketches was concerned, the cobra was regarded as the phallus."

No such pictures were in evidence at Shirala, but what was this, then, if not the men getting into the spirit of Nag Panchami and having a symbolic, priapic go of their own? Getting it up by proxy, then measuring their best erections, just as most schoolboys will no doubt have done, if perhaps with a mere six-inch ruler? Competing for that ultimate male accolade, the biggest one in the class? Certainly, winning seemed paramount, so much so that when the winner was announced, an incensed rival set about him with a measuring stick which he had just broken across his knee. Within seconds, other competitors were weighing in while the suits flicked fingers at the microphone in a vain attempt to repair it and so restore order.

"Oh dear," said the voice at my shoulder. It was Sayga Ghatge. "I think the wrong club won."

"Club?" I asked.

In the run-up to the festival, Sayga explained, family and friends got together to collect cobras from the wild, to take part in the various cobra competitions and to display the cobras in the parades. Some such snake clubs, such as the Rotary Club and the Lions Club of Shirala, were created from existing associations. Others, formed specially for Nag Panchami, included the Nagraja Trimurthi (the Snake King Trinity Club), the Hooded Cobra Club and even the advertising-aware Azzad Restaurant Club.

"And we," said Sayga, "are called the Young Stars." I asked why they had chosen that name.

"Because we are young," Sayga replied with certainty. "And we are stars."

Sayga took my hand; I was his guest, he said. And so the Young Stars chanted their way through town, a purple procession congregated around a bagged cobra hanging from a stick. The arcaded shops opposite Tirupati Auto Parts that usually sold fluorescent tubes, musical doorbells and scooter horns were being transformed for the festival into carpeted studios, with Kodak film cartons and balloons hanging by strings from the ceiling and a moth-eaten purple backdrop, where you could have your photo taken for a fee with a *dhaman*, a long, brown ratsnake.

"Not poisonous today, this snake!" the proprietor gestured urgently, inviting me in. The dhaman was draped around his neck, like a scarf. "Poisonous only on Sundays, the dhaman."

Where the arcade stretched downhill, it became increasingly gap-toothed as patches of open ground snagged between the shops. The countryside began reasserting itself, and beyond the lumber yard, the Ghatge family business where they made tomato crates, Shirala's main street finally became the road to Kolhapur. It ran along the edge of the playground, and disappeared past the courthouse down a funnel of great vine-tangled banyan trees. The Young Stars detoured along a dirt track which ran towards the temple and the town hospital, past stalls in various stages of assembly selling coconuts and bananas, pulses and spices, bracelets and balloons, rosettes and pictures of famous scenes from the Hindu epics, stalls smelling of camp fires and coriander.

Below the track an expanse of cropped grass, the unofficial village green, unfolded. Sections of metal and rolled up awnings with frills were being unloaded from brightly painted lorries. A fairground was taking shape against a rain-laden afternoon sky. Engines spluttered and miraculous peals of music briefly escaped from the tangle of loose connections. Aged carousels were being assembled. A series of crudely welded, lime-green helicopters, Bell Hueys perhaps, lay strewn across the ground like a scene from the Tet Offensive. The Well of Death—"Silver Enterprise presents Maruti Car with Motar-Cycles"—was rising from the ground, like a giant bucket thicketed in wood scaffolding. On a half-constructed stand, a woman's brightly painted image half voluptuous disco diva, half hair-in-bun grandmother—took form in random wooden sections. It was not until the last one was forced into place, with a determined hammering and not a little splintering (the stand was on an awkward slope), that the head of a great python, yellow and black, finally explained the lashing coils in which the painted lady was swathed.

The Ghatge home was a low stone building that trailed wood smoke from the windows. It was full of Ghatges. Old men, excited children and young women wrapped in neat saris gathered for a sighting of the Young Stars' latest cobra catch. Sayga's gangly and lugubrious cousin Vinod stepped forward. Vinod, who had masses of curly hair, dark, faintly troubled eyes and who was by all accounts the Young Stars' leading snake man, undid the bag and shook the cobra free. As the cobra emerged, Vinod took up an empty clay pot which he waved before the rearing snake, using it variously as red rag and protective shield. From the smoky shadows of the Ghatge home, several women stepped forward. They took pinches of burnt rice, hibiscus petals and turmeric from the tin trays they held, leaned forward and sprinkled them upon the cobra's head before genuflecting slightly and stepping back.

These women were worshipping a cobra. The fact was worth emphasizing. But they did so with a striking serenity. About their actions there was an absolute absence of self-consciousness. They worshipped with dignity and without question. But for this, their behaviour might have struck me as plain weird, but they betrayed no suggestion that worshipping cobras might seem inappropriate or obscure behaviour in the late-twentieth century. In their relationship with cobras, they were free from doubt.

The Ghatge women also demonstrated that females were the true Nag Panchami protagonists and the main players in the myths that informed it, whatever the representative phalluses on Ubale Lane might have indicated. The women worshipped the cobra as the procreative genius and as rat-consuming, rain-bringing friend of farmers, but they also placated its angered spirit. In Shirala, failing to produce children was widely regarded as divine punishment for having killed a cobra in a previous life. Of the several parables that informed Nag Panchami, the people of Shirala were particularly attentive to that of the villager who once sliced up a clutch of young cobras under his plough. The mother cobra, discovering the blood of her dead babies smeared upon the villager's plough, bit him and his family in a fit of night-time vengeance, killing all before they awoke. The cobra then made for a neighbouring village to seek out the one survivor of the family, a daughter who had married and moved away. But the cobra found her making offerings of frankincense, milk and honey to the snake gods and so forgave her, so much so that she even gave the daughter some nectar that she might sprinkle on her family to restore them to life.

The people of Shirala took especial care not to harm cobras during Nag Panchami. And they made a point never to turn the earth during the festival. The ploughs stood still, and the tractors were driven from the fields to pull the trailer floats in the festival parade through the village.

"And you will join the Young Stars on our trailer tomorrow!" said Sayga. When I asked Sayga what tomorrow's itinerary was, he scribbed it upon a piece of paper: *Dawn*, it read. *Snaks to temples. Eleven o'clock; snaks to houses. Lunchtime; snaks parade.*

I bade the Ghatges good evening and walked through Shirala in the gathering dusk. Rain clouds were hanging low over the town like udders, but through the gaps between them sharp stars pricked the sky. Back at my room, Mr. Patil had swept the onions and garlics from a neat rectangle of floor that was just large enough to accommodate the thin mat he had provided. A pathway of sorts had been cleared between the mat and the door and onwards to the standing tap at the back of the room, but it was continually being overrun by subsiding bulbs that would settle around me in the course of the night.

Mr. Patil's mat was not comfortable. Cacophonous night noises queued up for my attention; dogs howled, horns hooted, pigs squealed, onions and garlics tumbled, and battalions of televisions did battle, music and urgent dialogue sounding from them. The stonemasons worked late and in the middle of the night a frenzied hammering upon the water pipes began somewhere in the building. An Italian aria even sounded out. It then started to rain, and when fleets of buses began arriving from Bombay, Poona and Solapur in the early hours, unloading their pilgrim passengers on the school cricket pitch, the texture of sound became so laden it seemed chunks of it might finally fall from the sky to flatten the ragged tents or threaten the neat lines of Mr. Patil's apartment block.

But it was the sound of brass bands that awoke me. I dressed quickly and followed the dawn crowds that merged, like tributaries, with the main procession as it poured through town, sluiced past Youth Federation Corner and the hospital and gathered strength as it made for the temple, a tumble of motion, colour and chaos, its banks lined with the outstretched offerings of hawkers and stallholders. There were men in white Nehru caps, with the gold and red tridents of Shiva painted on their foreheads; the old men with white moustaches and orange turbans; the women in their gracefully cylindrical tangerine saris, the bridges of their noses daubed scarlet and orange; the orange flags topped with golden tridents, and the monitor lizards, another local reptile revered for its associates with Shiva, trussed to vertical poles and carried like the standards of Roman legionaries; the brass bands, ragged but splendid, kitted out in crepe, gold braid and Bacofoil—like Michael Jacksons created in thrifty home economics classes—marching beside the little roofed carts to which pewter megaphones, their amplification systems, were lashed. The New Hanuman Musical Band, The Durbar Brass Extravaganza or Musical Banjo, Miraj, as they called themselves, played trumpets, bugles, tubas, clarinets, drums and, most impressively, huge sousaphones like oversized French horns that largely obscured the players beneath them; the day's first purple explosions of gulal caking hair, faces and freshly laundered light blue shirts; and everywhere, borne along on shoulders as if bobbing upon the surface of the current, hundreds of brightly wrapped pots carrying the festival's true stars.

As the temple drew near, the stream slowed to a shuffle. Hawkers pressed bags of nuts upon us and aimed their paint-dipped metal trident stamps at undecorated foreheads in the expectation of a few rupees. An albino dwarf lay naked in the dirt, a single shiny coin beside him. Before him, the ivory-

coloured temple, mildewed with long stains of black, arose
from the high-walled courtyard that enclosed it. It was a square
structure backed by a tower that spired to a decorative pinnacle
which was shaped like a pineapple. A motif depicting the
hooded cobra was repeated all around the temple tower, half-
way up. The shredded orange prayer flags fluttered from the
roof.

The crush gradually extruded pilgrims into the courtyard
where they slipped off their shoes and laid their hands upon the
black statue of Nandi, Shiva's bull protector. They rang the
bells to announce their coming and made their way to break
auspicious coconuts at the shrine of Ambabai, as the consort of
Shiva is known in this part of Maharashtra. The men with the
pots released their cobras in a corner of the courtyard, and a
crowd gathered. The women abased themselves, then stepped
forward with their steel trays. The cobras did not move, maintaining the perfect curves of their taut hoods through the
shower of jasmine petals, turmeric, gulal and parched rice, as if
they were due such worship and understood what made them
magnificent.

Later that morning I returned to the onion store to sleep,
and dreamed of purple cobras in bags and pots, and rat-snakes
slithering over photo orders, and photos of pilgrims formally
posed in the Indian way, a little correct, but festooned with
snakes. I awoke to a knock at the door. Mr. Patil was there.

"Ah," he said. He had brought two men with him who carried
a clay pot.

"They have brought a nag for you to worship," said Mr. Patil.
"It is the tradition." Sayga's scribbled itinerary, snaks to houses,
now made perfect sense. I rolled up my bedding as best I could,
swept my travelling detritus to one side and calmed myself with
thoughts of all the surrounding garlic. Space was limited; when
the cobra emerged it was competing for space with my copy of
Vikram Seth's *A Suitable Boy*, a can of mosquito repellent,
some unsavoury underwear, a toothbrush and a great many

onions. But the domestic context only enhanced the snake's performance. After a few minutes, the men potted the cobra and accepted some rupees. They then asked me whether I would join them on their float for the afternoon parade. I was already taken, I told them; a guest of the Young Stars. I closed the door behind them and leaned against it, breathing deeply. *I just had a cobra in my room.* A real one, not one from my dreams. I was making progress.

By early afternoon, tens of thousands of people had descended upon Shirala. I made my way through the crowds to the main street, as Sayga had directed, and found the Young Stars parade trailer yoked to a tractor and preparing to join the parade. The Young Stars were perched along the sides of the trailer. I took my place among them and looked around me. A number of cobra pots lay at the front beneath a display table. The Young Stars were clearly enjoying themselves; their excited eyes stared out at me from their purple faces. The mood was high and their float was magnificently decorated. It was festooned with coconuts and palm fronds, and cocooned in an elaborate wooden structure that rose precariously above us, like a temple tower, snagging alarmingly on the phone lines and electricity cables that criss-crossed the street.

"Constructed in our own lumber yard," shouted Sayga proudly, pointing upwards.

As the trailers pulled out onto the street, Vinod began getting the snakes out. He placed them, one by one, upon the display table and his eyes seemed to flash as the cobras reared up. Immediately behind us, The Royal Musical Band Company hammered out its inimitable version of "When All the Saints" and the crowd broke into frenzied dancing. Families watched from the bulging balconies that overhung the street. Crowds milled beneath us, a stream of faces crossed by joy and occasional fear as the crush increased to an alarming level. Occasionally, anxious fathers would pass children up to us, to us of all places, as if the safest place for children in this bizarre version of the world was a trailer full of cobras. Bewitched, the children would stare at the snakes just a few feet away from them, and would be staring even as their fathers retrieved them some time later, scooping them up with an outstretched arm and a shout of thanks.

The atmosphere, I guessed, had begun to get to all of us, not least to Vinod. Vinod spent more time working the cobras, holding the pots before them or securing the snakes from behind, than any other Young Star. His face retained that same concentrated intensity but there was also a rapt, trance-like quality to his features now, as if this was his day in the sun and he was capable of anything. Vinod began cutting corners. At the beginning, he had held each cobra by the tail while another waved the pot at it. Now, emboldened, he supported them along his arm where they reared to face the pot. When he had finished, he grabbed the cobras with a reckless nonchalance, as if his skills made him immune to them.

The crush had increased. The sousaphone of The Royal Musical Band Company was pushed hard up against the back of the trailer, its gaping horn open like a huge flower. The noise was almost unbearable. One of the Young Stars, who had been flinging gulal at the passing balconies, released a handful down the horn of the sousaphone and everybody laughed as the great instrument fell silent. Then, with a great gust of breath the musician cleared the sousaphone, enveloping the trailer in a dense purple cloud which everybody, and the cobra on display, was unprepared for. As its world turned purple, the cobra wheeled away and bit Vinod viciously on the forefinger.

It had happened, what I'd anticipated for the best part of my life, and it had not happened to me. The bite had fallen elsewhere, and with it came the blissful, liberating sense that I must now worry not for myself but for somebody else. Vinod was clutching his finger where a tiny jewel of blood had appeared. The colour might have drained from his face; through the gulal, you couldn't tell. But the rapt expression had certainly gone, and a look of injured surprise had taken its place. To mark the sousaphone's recovery, the band played loud as ever. Initially, only the Young Stars realized what had happened. One returned the cobra to its pot. Several others helped Vinod from the trailer and made a path for him through the crowd to the nearby hospital. I followed.

Vinod was unsteady by the time he reached the hospital, a purple Young Star at each elbow. Walking along the green hospital corridor, his legs crossed like a drunkard's. When they sat him on a bed, he was slurring his words. Vinod was seriously envenomated.

The Young Stars were permitted to see him later that evening. But he was in no condition to tell them how he felt. He lay in a deep, childlike sleep, wrapped in a cocoon of blankets. A light sheen of sweat patterned by a purple marbling of sludgy gulal lay across his face. Young Stars and Ghatges alike perched on the edge of his bed and on chairs around the edge of the room, shadowy purple figures whispering among themselves like bereaved party-goers whom misfortune had struck at the height of their happiness. At the foot of Vinod's bed lay five empty vials of antivenom.

20. THE SNAKEBITE SURVIVORS' CLUB

The door handle turned. The hinges creaked. Darlene crawled outside. She pushed the door gently behind her, as if it might keep him inside, and gulped at the night air. It was fresh and reviving, and it brought her unsteadily to her feet. The night was starlit, but her eyesight was blurred and she moved by instinct and by feel, palming her way along the wall with her good hand, away from Glenn, away from the snake shed and all the terrible things that had happened to her in that house. Her left hand was so painful that she could not use it; like a broken wing, she dangled the arm clear of her body.

When she reached the garage, she took a deep breath and struck out unaided on the diagonal for the dry creek bed and the wire gate beyond, where the property ended and Barbee Lane began. A loud noise sounded in her head. It was full of strange sounds, of falling leaves dry as tinsel and running water, as if the trees and the river were whispering to her, urging her onwards. She had left him. But she was not safe yet. Any moment he might yet awake, his head muzzy with drink, find her gone and reach for the pistol. She stumbled on the track and retched emptily, swallowing back the noise. She looked back at the house, a blur of curtained lights, and thought for a moment that she could see him coming after her, to drag her back to the shed. She got to her knees and levered herself upright. She passed through the wire gate and began to move down Barbee Lane, flanked by fields stretching far beyond her broken vision.

She realized that she was fleeing, heading down Barbee Lane towards an approaching ambulance. Now, she could almost see him awakening without her. He would open his eyes in the morning, his mouth foul with drink and cigarettes, and look across the room to the pitiful corpse he had planned for. And she would not be there.

And then the ambulance was pulling up nearby, and a man was approaching her. She wanted to be taken to the hospital, she told him. The paramedic was ushering her into the warmth of the ambulance interior, which smelt clean and scrubbed and was full of steady voices talking on the two-way radio. And Glenn Summerford slept on, unaware that he was looking at ninety-nine years in a state penitentiary (minus, as the judge allowed, the twenty-one days he had already served inside). The paramedic was bringing her water, washing her hands and urging her to rest. Then she was quiet and he settled back to fill in the pre-hospital care report. Under the heading "chief complaint," he wrote the words "Snakebite x 2" and, as the ambulance driver put in several tight turns against Barbee Lane's narrow verges and drove through Scottsboro to the county hospital, the paramedic turned to tell her that she was OK now.

OCTOBER 1996

On my last day in FNQ, a headline in the *Cairnes Post* caught my eye. "Woman injured in cassowary attack," it read. A woman walking on a track in the Aeroglen area of Cairns had suffered two gashes to the head requiring nine stitches and two punctures to the left thigh, when a large bird about the size of an ostrich had attacked her. Appalled, I read on. The newspaper quoted the woman,

> I heard noises and turned round to see a cassowary fifteen metres behind me. Soon it started to gain speed on me...I stepped aside and waited on the edge of the trail but it just came up and stood beside me. We eyed each other for about

fifteen seconds then it just whacked me and threw me to the ground. It hit me again and I could feel blood on my head. I was lying on the ground, pretending not to move when the cassowary struck me again, kicking me a distance downhill.

If I understood this right, a woman had been attacked by a giant bird. Even by the standards of FNQ, where the wildlife seemed to have endangered the most routine activities, making even a walk in the woods a risk to life and limb, this was going it some. You couldn't swim in the sea for the box jellyfish, the rivers for the crocodiles. Then, if you survived them and the spiders and all the other creepy-crawlies, and if the dive-bombing magpies or the aggressive seagulls didn't get you, then you still had bloody cassowaries to worry about. And taipans. The whole business, which had begun by scaring me, was now beginning to irritate me. And even before I'd finished reading about the cassowary attack, I knew that I had to go back to Mareeba.

OK, I'd endured the pet shop taipans. Back at his house, Brian Starkey had even got a taipan of his own out of its cage for me (although I don't mean to suggest I did anything more than stare stupidly at it as Brian kept a grip on it with his grab stick). But I knew that wasn't enough. What scared me was walking the Barron River. Which was why I had to spend my last day going back there, and walking it.

I owed it to myself and to Browne Hayes, to Gordon Sinclair, to Juanita in Seattle and the taxi-driver in Nairobi, and all the others who'd travelled with me and spent too much time scared of snakes. I had to walk that riverside path which, despite the old couple with flowers behind their ears and biking kids and all the common sense I could muster, had become a taipan-infested deathtrap in my anxious mind. I had begun to realize it was my mind, not snakes, I was really up against.

So I packed my boots, patted my pocket to check for my

ash charm and drove up from Cairns in the late morning. I parked at the foot of Mareeba's deserted Basalt Street where the walking track began. As I put my heavy leather boots on, heaving the laces in a tight criss-cross, I went through the walk in my mind; from the car, I would descend past the belvedere where the kids scrawled graffiti and came for clandestine smokes, down to the old brewery swimming hole where they used to hold the swimming carnivals until the new pool got built over by the rugby pitch, and I would be on the banks of the Barron. From there, the path ran for about a mile, I guessed, past the Herberton Street bridge, the bottom of Hastie Street, Atherton Street and, finally, Middlemiss Street where I could return to the car through town. Easy.

I smoked a cigarette, stamped the butt beneath my heel and set off down the track. I soon passed the belvedere and followed the steep incline down to the river. Initially, the ground either side of the concrete path was rocky and largely open; this gave me a good view of anything coming and progress was smooth. When I got down to the river level and the old brick-lined swimming hole where the Barron gathered, things crowded in a bit. There were fronds and things nibbling at the sides of the path. So I proceeded slowly and stamped my boots. My eyes scanned the path and the ground on either side. I checked my feet and the foliage. Nothing. I was listening for the noise I dreaded, that distinctive liquid slither, but all I heard were insects, the innocent gurgle of the river and the cars on the Herberton Street bridge. Later, I could hear kids playing on Hastie Street and a dog barking somewhere. I could feel rivulets of sweat sidewinding down my neck.

Then I was arriving at a place I recognized, and the path leading up to Middlemiss Street ran off to my left. I punched the air. I had made it. I had not been attacked. The feeling was so good that I did not take the Middlemiss Street track. Instead, I turned and retraced my steps. It struck me, on the way back to the car, that the Barron was a beautiful river.

NOVEMBER 1996

"May I speak to Pete Bramwell please?"

"You're speaking to him."

"The Pete Bramwell who got bitten by a black mamba in Kilifi, Kenya?"

"The same."

"I've been looking for you."

"You have?"

"I wanted to hear your snakebite story. I heard about it all over Kenya back in the summer...the bite, being put on the respirator, the fact you could hear everything they said but you couldn't move a..."

"The what?"

"You being paralyzed, I mean."

"Hold on. That wasn't me. You're getting me confused."

"That wasn't you?"

"No, that was the other bloke down in South Africa. Happened, what, a month or so before I got bitten. Our stories do tend to get confused. For some reason, his bite got written up everywhere, even in *Newsweek*. Come up to Banchory and we'll talk. Meantime let me send you the cuttings."

The cuttings arrived in due course. The other black mamba bite had happened at a snake park near Pretoria on March 14, 1977. The park owner was technically dead when they put him on life-support at Pretoria General Hospital an hour after being bitten. Later that day, he regained a kind of consciousness. He could hear his own breathing but he was absolutely paralyzed. "Even though I was unable to move or communicate in any way, I could hear everything around me and was totally aware of what was going on," he later explained.

In the days that followed, he heard specialists at his bedside discussing whether he had suffered brain damage between the time his heart stopped beating and being placed on life-support, and nurses denigrating the size of his penis as they put a catheter into it. He heard the doctor telling his wife they weren't sure that he'd recover consciousness, but if he did he'd be brain damaged. He heard his wife telling him, despite the doctors, how the family was and what was happening at the snake park. On the third day, he heard one doctor suggest the best thing they could do for him and his family was to disconnect him. He heard them finally decide to leave the machine on because his case was clinically interesting. That was the point at which he would have breathed a sigh of relief. If he'd been able to.

On the eighth day, he finally got a finger to twitch. A nurse noticed and called the doctor. The doctor spoke to him, asking him to twitch the finger again if he could hear him. The finger moved. The man was alive. And the paralysis was wearing off. They would leave him on life-support as long as he needed it. He would make a full recovery.

Two mamba bites had merged to create a story that people recounted in hushed tones twenty years later. The incidents had occurred just six weeks apart, so some confusion was understandable, especially since two white snake men were involved. But there was also a sense of design, of intention, behind the adaptations in the hybrid narrative, as if its many tellers were bent on creating a nightmare that would throw a dark but thrilling shadow over their lives. Bramwell's bite circumstances provided the necessary local setting but the South African's experience supplied most of the memorably horrifying narrative. Over the years, the story had been fitted, as stories tend to be, with several poetic improvements; the version I'd heard in Kenya had the South African's wife noticing the single tear at her husband's eye as the priest performs the last rites before the life-support is turned off (when it was in fact a nurse that noticed; what he'd actually done was move a rather more prosaic finger; and it never got as far as last rites), and Pete Bramwell blowing away all his mambas with a shotgun (when all he'd shot, as he explained, were any black mambas that came too near the house in the first few weeks after the incident). Their combined story had acquired the status of legend. Amended here, edited there, buffed and polished in the telling, it had come to encapsulate and accentuate the black mamba's natural qualities; as the stuff of nightmares.

Finally, there was the matter of the South African's name. He was called Jack Seale. Which sounded a bit like my name, and Jack Seale had just happened to be my age when the bite occurred. Strange, I suppose, but nothing like as strange as if I'd still been on Lariam when I can only guess at the thoughts this passing coincidence would have prompted; that if you allowed Jack Seale just the initial of his Christian name and if you only spoke the words, then you had my name. And my age. Me, in short. That I'd somehow been transported into a parallel universe where I had been bitten by a black mamba, just as I'd known I always would be. That I'd felt the sensations I'd always expected, the sudden seething, like a wave breaking brown and olive about my feet, the brief pain and the blood and venom seeping scarlet and yellow from the puncture marks as the snake flicked its tail in the long grass. That I'd died but been kept alive to lie in some hospital bed, my life hitched to a switch, listening to doctors who were sure I was already dead.

Like I say: too much imagination.

AUGUST 1996

On the morning after Nag Panchami, unlikely English sentences floated up to me on my mat in the onion store.

"A cloudless sky in the month of Shravan can really worry the poor Indian farmer," the strange voice intoned. "Astronauts remain weightless during travel," it continued. I looked over the balcony. It was C.Y.W., the postman, taking a break from his delivery round to read from his English phrasebook. Then I remembered what had happened to Vinod and I dressed quickly.

"No, that is a lame excuse for overcharging," said C.Y.W. in his best affronted shopper's voice, smiling at me as I passed. I nodded encouragingly and hurried through the mud towards the hospital. Vinod was standing outside. He had just discharged himself. He looked as gangly and cadaverous as ever, but a little sheepish too. He led me to understand that he was feeling OK, just a little groggy. The antivenom had done its work.

He took my arm. They were about to release the cobras, he explained. The Young Stars were gathered at the Ghatge house, piling the cobra pots onto the tractor trailer. When they had finished, a dozen hands helped Vinod climb gingerly onto the trailer. Then we were trundling down a dirt track in the shade of the wooden temple structure that had somehow survived the celebrations.

"Once people heard about Vinod," Sayga explained, "nobody really felt like smashing it up, like we normally do."

The track led past a final few houses. The tractor drew up in the corner of an uncultivated field. The Young Stars lifted their pots and clambered from the trailer. Vinod insisted on carrying the cobra which had bitten him. A few yards into the field, they undid the pots and gently tipped out the cobras. The snakes, still purple with gulal, reared as they emerged but finding their way unhindered, streaked towards the undergrowth.

As we watched them go, Sayga spoke to me, telling me that the Young Stars would be back to collect more cobras next year. I turned to him and smiled, and when I looked round again, the snakes had disappeared.

NOTES

Subsequent references to publications are indicated using the author's surname only.

PROLOGUE

PAGE 2 *"known many women..."*: Beatrice Thompson, *Who's Who at the Zoo*, p. 105.

1. THE REPTILE HOUSE

The following sources supplied information on the history of London Zoo's reptile house: Wilfred Blunt, *The Ark in the Park* (Chapter 20: "Snake Troubles"); P. Chalmers Mitchell, *Centenary History of the Zoological Society of London*; H. Scherren, *The Zoological Society of London*; and Catherine Hopley, *Snakes: Curiosities and Wonders of Serpent Life*.

PAGE 4 Epigraph: Enid Blyton, *The Zoo Book*, p. 69.

PAGE 4 *"interesting psychological entertainment"*: E. G. Boulenger, *A Naturalist at the Zoo*, p. 159.

PAGE 4 *"BEAUTIFUL RATTLESNAKE ALIVE"*: Hopley, p. 285.

PAGE 5 *"there was not much difficulty ..."*: Broderip's account is quoted in Scherren, pp. 92–3.

PAGE 5 *John Girling*: *The Times*, October 23, 1852, and Blunt, Chapter 20, "Snake Troubles."

2. AMERICA I

I am indebted to the following essential texts on the Appalachian snake-handling churches: Thomas Burton, *Serpent-Handling Believers*; Weston LaBarre, *They Shall Take Up Serpents*; Dennis Covington, *Salvation on Sand Mountain*; David Kimbrough, *Taking Up Serpents*; and Steven M. Kane's article "Appalachian Snake Handlers." I also referred to the *Atlanta Constitution*'s coverage of the Summerford trial (March, 1992), and to the trial transcript, *The State* versus *Summerford*, held at the Jackson County Courthouse, Scottsboro.

3. AUSTRALIA I

The main sources for the Browne Hayes story are Charles Henry Bertie, *The Story of Vaucluse House*; K. R. Cramp, *William Charles Wentworth of Wentworth House*; Suzanne Mourot, *Vaucluse House; Origin of this Historical Monument*; and John Lang, *Remarkable Convicts. Or Recollections of Botany Bay*.

PAGE 25 Epigraph: letter held at the Derbyshire Records Office, quoted in Robert Hughes, *The Fatal Shore*, p. 435.

PAGE 26 *"bad and filthy bedding…"*: Letter from Sir Jerome Fitzpatrick to Rev. Charles Lindsey, Pelham Papers, held at the British Library and quoted in Hughes, p. 149.

PAGES 26–7 *"On coming to…"*: Alexander Marjoribanks, *Travels in New South Wales*, p. 12.

PAGE 27 *"What the bushman…"*: Horace Wheelwright, *Bush Wanderings of a Naturalist: Or Notes on the Field Sports and Fauna of Australia Felix*, p. 197.

PAGE 27 *"What To Do …"*: Reverend John Flynn, *The Bushman's Companion*, under "What To Do In Case Of Snakebite."

PAGE 27 *"the queer matrimonial speculation"*: Paper by Morgan McMahon, 1914, and quoted in Bertie, p. 2.

PAGE 29 *"rather fresh-coloured…"*: Bertie, p. 3.

PAGE 29 *"special convict"*: Lang, p. 134.

PAGE 29 *"flower and seed list"*: Bertie, pp. 10–11.

PAGE 30 *"REWARD OF TEN GUINEAS…"*: *Sydney Gazette* advertisement, July 21, 1804.

PAGE 30 *"TEN GUINEAS REWARD"*: *Sydney Gazette* advertisement, February 9, 1811.

PAGE 30 *"The part of the country…"*: Lang, p. 137.

PAGE 30 *"fine boy"*: *Sydney Gazette*, October 14, 1804.

PAGE 31 *"Now, so long as…"* and *"I have been praying"*: Lang, pp. 137–8.

PAGES 32–33 *workmen uncovered* … : The discovery of the Hayes ditch is recounted by Bertie, p. 24.

PAGE 33 *"'Take it,' Sir Henry replied…"*: Bertie, p. 24.

PAGE 34 *vipers were traditionally dispatched* …: Daphne Du Maurier, *Vanishing Cornwall*, pp. 9–10.

4. AFRICA I

PAGES 36–41 For much of my material on the Uganda Railway I am grateful to Ronald Hardy, *The Iron Snake*, and J. H. Patterson, *The Maneaters of Tsavo*.

PAGE 41 *"rushed to the tent..."*: Patterson, p. 164.

PAGE 42 *Amam Din*: Hardy, p. 211.

PAGE 46 *"saw a large snake..."*: W. C. Willoughby, *The Soul of the Bantu*, p. 166.

PAGES 46–7 Mr. *Henry*: A. Henry, *Travels and Adventures in Canada and the Indian Territories*.

PAGE 47 *"If such a snake..."*: Reverend H. Callaway, *The Religious System of the Amazulu*, p. 107.

PAGE 47 *"Divine snakes enter a house..."*: Callaway, p. 107.

PAGE 50 *"The diabolical ingenuity of Nature..."*: F. W. Fitzsimons, *Snakes*, quoted in Charles R. S. Pitman, *The Snakes of Uganda*, pp. 237–8.

PAGE 50 *"I had seen the menace..."*: Eugene Marais, *The Road to Waterberg*, pp. 71–77.

5. AMERICA II

PAGES 68–9 *"A multitude of spots..."*: Hector St. John De Crevecoeur, *Letters from an American Farmer*, p. 175.

PAGE 75 *"Often at night..."*: Elijah Bingham, *Snake Stories from Africa*, pp. 10 and 19.

PAGE 77 *"the sight of a hedious serpent..."*: "Stirling," *Wonderful Account of the Rattlesnake and Other Serpents of America*, pp. 5 and 9.

PAGE 77 *"the atmosphere which surrounds it..."*. Hopley, p. 153.

PAGE 77 *"woodsmen...are in a fright..."*: Walduck is quoted in Hopley, p. 153.

PAGES 77–8 *"Yea there are some Serpents..."*: Reverend Francis Higgeson, *New-England's Plantation; or a Short and True Description of the Commodities and Discommodities of that Country*, quoted in Klauber, p. 311.

PAGE 78 *"It is certain..."*: "Stirling," p. 3.

PAGE 78 *"went to mowing"*: De Crevecoeur, pp. 176–7.

PAGE 78 *"blew, white and greene spotted"*: Klauber, p. 311.

PAGES 78–9 *"It is simplicity..."*: Thomas Morton, *New English Canaan* quoted in C. Curran and C. Kauffeld, *Snakes and their Ways*, p. 46.

PAGE 79 *"may be esteemed an emblem of vigilance..."*: *Pennsylvania Journal*, December 27, 1775. Quoted in Harry W. Greene, *Snakes: The Evolution of Mystery in Nature*, p. 117.

PAGE 79 *"The colours of the American fleet..."*: *London Chronicle*, July 27, 1776. Quoted in Curran and Kauffeld, p. 222.

6. AUSTRALIA II

My main source for the history of Cairns is Dorothy Jones, *Trinity Phoenix: A History of Cairns and District*.

PAGE 82 *"a splendid and unequalled Marine Estate..."*: Jan Morris, *Sydney*, p. 66.

PAGE 83 *"the most danger..."*: John Lawson, *A History of Carolina*, quoted in Hopley, p. 283.

PAGE 83 *"We had not gone far..."*: J. Huxley (ed.), *T. H. Huxley's Diary of the Voyage of H.M.S. Rattlesnake*, p. 281.

PAGE 84 *"land the treasure there"*: J. MacGillivray, *Narrative of the Voyage of H.M.S. Rattlesnake*, vol. 1, p. 3.

PAGE 84 *"determine which was the best..."*: MacGillivray, vol. 1, p. 2.

PAGE 85 *"butterflies of great size..."*: MacGillivray, vol. 1, p. 85.

PAGE 85 *"Snakes...require to be carefully avoided..."*: MacGillivray, vol. 1, p. 130.

PAGE 85 *"wide creek running..."*: MacGillivray, vol. 1, p. 99.

PAGE 89 *"twelve people..."*: *Australasian Medical Gazette*, 1890. Vol. 9, p. 157.

PAGE 89 *Robert Crabbe*: *Cairns Post*, October 14, 1920.

PAGES 90–1 *orange-bellied brown snake* ... : Gerard Krefft, *The Snakes of Australia*, contents pp. IX–XVI.

PAGES 94–6 *Kevin Budden*. I am indebted to David Williams, "The Death of Kevin Budden" (unpublished paper), and to David Fleay, "Adventures with a Taipan," *Animal Kingdom*, vol. 56, 1953.

PAGE 96 *"six foot five inches long..."* and *"dealt with the most savage..."*: Fleay.

7. AFRICA II

PAGE 101 Epigraph: Isak Dinesen, *Out of Africa*, p. 214.

PAGES 101–2 *"noted for its many poisonous snakes"*: William Walter Augustine Fitzgerald, *Travels in the Coastlands of British East Africa*, pp. 146.

PAGE 102 *"in a ghastly funk"*: Fitzgerald, p. 196.

PAGES 102–3 *"This half-way camp..."*: Fitzgerald, pp. 195–6.

PAGE 115 *"a trembling fit..."*: Fitzgerald, p. 268.

PAGE 115 *"dragging a puff adder..."*: Fitzgerald, pp. 267–8.

PAGES 117–8 *Richard Meinertzhagen*. Richard Meinertzhagen, *Kenya Diary 1902–1906*, p. 321. The incident is also recounted in James A. Oliver, *Snakes in Fact and Fiction*, pp. 76–77.

PAGES 118–9 *Dr. F. Eigenberger*: F. Eigenberger, "Some Clinical Observations on the Action of Mamba Venom," *Bulletin of the Antivenin Institute of America*, June 1928, p. 45.

PAGE 121 *the Luo*. Pitman, page 243. Lawrence Green, *Secret Africa*, claims that the Ndorobo people also tied deadly snakes near pathways.

8. AMERICA III

PAGES 123–4 *"the rattlesnake hazard..."*: Klauber, p. 180.

PAGE 124 *"snake-hunting has been a pastime..."*: Krefft, p. 6.

PAGE 130 *Back in 1849, two Iowa men*: Klauber, p. 235.

PAGES 133–4 *"hedious serpent..."* and *"by the rapidity of its motion..."*: "Stirling," p. 3.

PAGES 139–40 *Dolly Dickens*: J. Frank Dobie, *Rattlesnakes*, pp. 87–90.

9. AUSTRALIA III

My main reference sources on the Australian snake show tradition are John Cann, *Snakes Alive!*, the introduction to Struan Sutherland, *Australian Animal Toxins*, and W. Crowther, "Mr. Charles Underwood and his Antidote, with some Observations on Snake-Bite in Tasmania" in *The Medical Journal of Australia*, January 21, 1956.

PAGE 144 Epigraph: Krefft, p. 15.

PAGE 147 *"blasted with gunpowder"* and *"quite ill for eight hours..."*: *Australian Medical Journal*, vol. 4, pp. 81–96.

PAGE 147 *"Young Milo..."*: Cann, p. 75.

PAGE 148 *"cured without failure..."*: Cann, p. 61.

PAGE 148 *"discovered the secret..."*: Wheelwright, p. 202.

PAGE 149 *"the antidote vendors and their supporters..."*: Krefft, p. 14.

PAGE 150 *"The green mamba wins"*: Cann, p. 115.

PAGE 151 *"This case...presents several features..."*: *The Medical Journal of Australia*, March 9, 1929, p. 307.

PAGE 152 *"Show Time is African Pigmy Time"* and *"See the pigs perform..."*: *Cairns Post*, July 1951. Various dates.

PAGES 152–3 *Paula, the Exotic Snake Dancer* and *Melinda Lee*: *Cairns Post*, July 17, 1951.

PAGE 155 *"More Fatal Cases..."*: *Medical Journal of Australia*, October 7, 1944.

PAGE 157 *"an enormous disturbance..."*: *Cairns Post*, December 23, 1985.

10. AFRICA III

PAGES 163–4 *the mchowis of the Wakamba* ... : C. W. Hobley, *Bantu Beliefs and Magic*, p. 199.

PAGES 164 *Their unrivalled snake dawa*: F. G. Carnochan and H. C. Adamson, *The Empire of the Snakes*; pp. 93–4.

11. INDIA I

The indispensable source on the classical cobra is J. Ph. Vogel, *Indian Serpent-Lore or the Nagas in Hindu Legend and Art*.

PAGE 168 Epigraph: Gordon Sinclair, *Foot-Loose in India*, p. 216.

PAGE 172 *"above all cattle..."*: Genesis 3, v. 14.

PAGE 172 *"He worshipped the nag"*: J. H. Rivett-Carnac, "The Snake Symbol in India." *Journal of the Asiatic Society of Bengal*, vol. XLVIII, part 1, 1879, p. 24.

PAGE 173 *"news chaser," "foot-loose prowler"* and *"he-man land of adventure."* Sinclair, frontispiece and p. 1.

PAGE 173 *"A swell lullaby..."*: Sinclair, pp. 261–2.

PAGE 173 *"Every seven minutes..."*: Sinclair, p. 161.

PAGE 173 *"casual item"* and *"Cobras killed three people..."*: Sinclair, pp. 133–4.

PAGE 174 *"He hadn't been bitten..."*: Sinclair, p. 133.

PAGE 174 *"In case you happen to be right in the cobra belt..."*: Sinclair, pp. 109–10.

PAGE 174 *"With the first peep of dawn..."*: Sinclair, pp. 58–60.

PAGE 175 *"one of those superstitious outcroppings..."*: Sinclair, p. 61.

PAGE 175 *"it was [Satan's] device..."*: Rev. John Bathurst Deane, *On the Worship of the Serpent*, p. 36.

PAGE 175 *"the little platform..."*: Rev. W. S. Durham, *Snake Worship and Other Topics*, pp. 1 and 3.

PAGE 177 *"there are certain obvious points..."*: Rivett-Carnac, p. 18.

12. AMERICA IV

PAGE 181 "*You will see and hear...*": trial transcript, *The State* versus *Summerford*.

PAGES 182–3 *the Scottsboro Boys*: Kwando M. Kinshasa, *The Man from Scottsboro; Clarence Norris in His Own Words.*

PAGE 184 *black churches...cross-burnings...bath towels...*: Georgia State Advisory Committee to United States Commission on Civil Rights, *Perceptions of Hate Group Activity in Georgia.*

13. AUSTRALIA IV

PAGE 195 *A Mossman cane farmer...*: Letter to author from Billy Brown and Clive Morton, January 31, 1997.

PAGE 196 "*Death by snakebite in Queensland...*": K. A. Blyth, "The Taipan—Truth and Treatment," *The Producers' Review*, Queensland Cane Growers' Council, October 1956.

PAGE 196 "*Too Much Talk of Taipans*": *Cairns Post*, July 27, 1956.

14. AFRICA IV

PAGE 208 Epigraph: Quoted in Russell Hope Robbins, *The Encyclopaedia of Witchcraft and Demonology*, p. 524.

PAGE 209 "*It suddenly occurred to me...*": John Crompton, *The Snake*, pp. 109–10.

PAGE 210 "*a pole standing upright...*": Willard Price, *Gorilla Adventure*, pp. 121–3.

PAGE 211 *Lariam's anagrammatical possibilities.* John Ryle, "City of Words," *Guardian*, April 14, 1997.

15. INDIA II

PAGE 224 Epigraph: Patrick Russell, *An Account of Indian Snakes Collected on the Coast of Coromandel.*

PAGE 227 *a noted charmer of Luxor, Egypt*: Curran & Kauffeld, pp. 154–5.

PAGE 227 "*take the precaution to excite the snake...*": Abbé J. A. Dubois, *Hindu Manners, Customs and Ceremonies*, p. 75.

PAGE 227 "*Here's a scrawny lad...*": Sinclair, p. 170.

PAGE 230 "*During my whole residence in India*": Dubois, p. 647.

PAGE 231 *Sir Joseph Fayrer*: Fayrer's snakebite researches, published in his *Thanatophidia*, are well excerpted in Brandt Aymar (ed.), *Treasury of Snake Lore*, p. 52.

PAGE 234 "*Cobra de Cappello*...": Russell.

PAGE 236 "*The fever belted jungle lands*...": Sinclair, p. 166.

PAGES 237–8 *Henry Sewall*: Greene, p. 78.

PAGE 238 "*not altogether discarded*...": Hopley, p. 532.

17. AUSTRALIA V

PAGE 261 "*already cyanosed and comatose*...": *Medical Journal of Australia*, July 6, 1940.

PAGE 261 "*enjoyed wide popularity*...": *Cairns Post*, October 24, 1938.

18. AFRICA V

PAGE 273 Epigraph: Peter Matthiessen, *The Tree Where Man Was Born*, p. 104.

PAGE 273 *William Fitzgerald wrapped it in inverted commas*: Fitzgerald, p. 172.

PAGE 274 "*a beautiful dark purple*...": C. J. P. Ionides, *Mambas and Maneaters*, pp. 155–6.

PAGE 280 "*an arrow from a bow*...": Marais, p. 74.

PAGES 280–1 "*It moves with great rapidity*..." and "*Periodically, it selects a haunt*...": Dr. J. O. Shircore, "Two Notes on the Crowing Crested Cobra," *African Affairs*, October 1944, pp. 183–6.

18. INDIA III

PAGE 291 and PAGES 298–9 *Nag Panchami*: Vogel, pp. 275–80, Vishwanath Mandlik, "Serpent Worship in India: The Nag Panchami Holiday as it is now Observed" in *Speeches and Writings*, and Harry Miller "The Cobra, India's Good Snake," *National Geographic*, September 1970.

PAGE 295 "*More than the usual license*...": Rivett-Carnac, p. 26.

20. THE SNAKEBITE SURVIVORS' CLUB

PAGES 306–7 "*Woman injured in cassowary attack*" and "*I heard noises*...": *Cairns Post*, April 3, 1996.

ACKNOWLEDGMENTS

I owe a special debt to the following books for the various, enduring impressions they made upon me: Dave Kimbrough's *Taking Up Serpents*; Dennis Covington's *Salvation on Sand Mountain*; Laurence Klauber's *Rattlesnakes*; *The Taipan* by Paul Masci and Philip Kendall; *Snakes Alive!* by John Cann; *Indian Serpent Lore* by Jean Vogel; Minton's *Venomous Reptiles*; and Curran and Kauffeld's *Snakes and Their Ways*. I am grateful to the following publishers for permission to quote from publications under their copyright: Macmillan Books for *The Maneaters of Tsavo* by J. H. Patterson and Kluwer Academic for *Travels in the Coastlands of British East Africa* by William Walter Augustine Fitzgerald. Every effort has been made to contact all copyright holders. I would be happy to rectify any omissions at the first opportunity.

I would also like to thank the large number of people who helped in the writing of this book by variously offering everything from advice, support, information, expertise, reminiscence and time, to accommodation and air tickets. Others alerted me to snake references, to programmes and articles I would otherwise doubtless have missed. In the U.K.: Peter Bramwell, Rahul Brijnath, Mary Coombs, Keith Corbett, Sophie Cottrell, Justin Creedy Smith, William Dalrymple, Jack Dillon, Susie Dunachie, Nick Fothergill, Jeremy Gavron, Robin Gauldie, Tony Gent, Navina Haidar, Heather Hall, John Hatt, Brian Jackman, Nigel Kotani, Fiona Mann, Edward Marriott, Cory McCracken, William McCracken, Richard Newton, Mark O'Shea, Bruce Palling, Ros and David Park, Ed Pugh, Janet Reynard, Veronica Ross, Candida Seal, Joanna Seal, Chris Sitwell, David Smethurst, Paul Sussman, Liz Tudball, Cath Urquhart, David Warrell, Sandy Wells and Wolfgang Wuster. I would also like to thank the helpful staff of the British Library, the Royal Zoological Society Library, the Royal Society of Medicine Library, and the Rhodes House and India Institute Libraries at the Bodleian, Oxford.

In the United States: Charles Allen and Kristen Rae, Darlene Collins, Jean Currie Church at the Moorland Springarn Research Center in Washington, Maynard Cox, Dr. David Hardy, Michele Lavery, John

Lodge, John Luckie, Dr. Sherman Minton, Tom Rhodes, Carrie Weeks at the Civil Rights Institute, Birmingham, Alabama, and Nanci Young at the Seeley G. Mudd Manuscript Library, Princeton, New Jersey. I would particularly like to thank Ben and Dorsey Jennings for all their hospitality in Atlanta, and Dave Kimbrough, Jimmy Morrow and Carl Porter.

In Australia: Jan Aland, Clive Brady, Bill Brown, Ken Blyth, Ram Chandra, Jeanette Covacevich at the Queensland Museum, Bill Kerr at Canegrowers, Peter Mirtschin at Venom Supplies, Clive Morton, John Nieuwenheuzen, Professor Richard Shine, Brian Starkey, Dr. Struan Sutherland, Darryl Teece, Dr. Julian White, David Williams, Aileen Park at the Cairns and District Family History Society, the staff of the Royal Historical Society of Queensland, and the *Cairns Post*. I would particularly like to thank Gwen Watson for her invaluable help. Thanks also to Virgin and Ansett Australia for help with flights.

In Africa: Jimmy and Sanda Ashe, Mike Bates, David Bristow, Dr. Donald Broadley, Ros Clark Percival, Will Craig, Iain Douglas Hamilton, Mark Easterbrook, Quentin Luke, Barbara Simpson, Jane Taylor, Jacob Undar, and Danny Woodley of the Kenya Wildlife Service. I would particularly like to thank John Pearman and Des Bowden, and Stefano Cheli, Liz Peacock and all at Cheli & Peacock Safaris for all their help and hospitality in Kenya. Thanks also to Nick Van Gruisen at Worldwide Journeys and Expeditions for kindly providing me with a flight to Kenya.

In India: Harry Miller, Atul Kale, Dr. Shantakumari at the King Institute, and Romulus Whitaker at the Madras Crocodile Bank. I would like to thank Cox and Kings for helping me with flight arrangements to India.

I would also like to thank the staff at Picador, especially Peter Straus, Richard Milner and Rachel Lockhart, and Jane Bradish-Ellames at Curtis Brown for all advice, support and kindness.

Lastly, love and thanks to Ash for pretty much everything, and to Anna, whose arrival proved the best kind of distraction over the last few months writing this book.

BIBLIOGRAPHY

GENERAL

Aymar, Brandt (ed.). *Treasury of Snake Lore*. Greenberg, New York, 1956.

Blunt, Wilfred. *The Ark in the Park*. Hamish Hamilton, London, 1976.

Boulenger, E. G. *A Naturalist at the Zoo*. Duckworth, London, 1926.

Chalmers, Mitchell P. *Centenary History of the Zoological Society of London*. Royal Zoological Society of London, 1929.

Crompton, John. *The Snake*. Faber & Faber, London, 1963.

Curran, C. and Kauffeld, C. *Snakes and Their Ways*. Harper & Row, New York, 1937.

Deane, Reverend John Bathurst. *On the Worship of the Serpent*. Hatchard & Son, London, 1830.

Ditmars, Raymond L. *Snakes of the World*. Macmillan, New York, 1937.

Du Maurier, Daphne. *Vanishing Cornwall*. Victor Gollancz, London, 1967.

Engelmann, W. E. *Snakes; Biology, Behaviour and Relationship to Man*. Croom Helm, London, 1984.

Frazer, J. G. *The Golden Bough; A Study of Magic and Religion*. Macmillan, London, 1922.

Godey, John. *The Snake*. Putnam, New York, 1978.

Greene, Harry W. *Snakes: The Evolution of Mystery in Nature*. University of California Press, Berkeley, 1997.

Hopley, Catherine C. *Snakes: Curiosities and Wonders of Serpent Life*. Griffifth & Farrar, London, 1882.

Howey, M. Oldfield. *The Encircled Serpent*. Rider & Co., London, 1928.

Minton, S. A. and M. R. *Venomous Reptiles*. Scribners, New York, 1969.

Morris, R. and Morris, D. *Men and Snakes*. Hutchinson, London, 1965.

Mundkur, Balaji. *The Cult of the Serpent*. University of New York Press, 1983.

Oldham, C. F. *The Sun and the Serpent*. Archibald Constable & Co., London, 1905.

Oliver, James A. *Snakes in Fact and Fiction*. Macmillan, London, 1958.

Scherren, H. *The Zoological Society of London.* Cassell & Co., London (Undated).

Scholefield, Alan. *Venom.* Heinemann, London, 1977.

Stidworthy, John. *Snakes of the World.* Hamlyn, London, 1969.

Thompson, L. Beatrice. *Who's Who at the Zoo.* Gay & Bird, London, 1902.

Wake, C. Staniland. *Serpent Worship and Other Essays.* George Redway, London, 1888.

AMERICA

Burton, Thomas. *Serpent-Handling Believers.* University of Tennessee Press, Knoxville, 1993.

Covington, Dennis. *Salvation on Sand Mountain.* Addison-Wesley, New York, 1995.

De Crevecoeur, J. H. St. John. *Letters from an American Farmer.* London, 1782.

Dobie, J. Frank. *Rattlesnakes.* Hammond, Hammond & Co., London, 1965.

Henry, A. *Travels and Adventures in Canada and the Indian Territories.* New York, 1809.

Higgeson, Rev. Francis. *New England's Plantation; or a Short and True Description of the Commodities and Discommodities of that Country.* London, 1630.

Kane, Steven M. "Appalachian Snake Handlers" in *Perspectives on the American South*, Vol. 4. Gordon & Breach, New York, 1987.

Kimbrough, David L. *Taking Up Serpents.* University of North Carolina, Chapel Hill, 1995.

Kinshasa, Kwando M. *The Man from Scottsboro; Clarence Norris in His Own Words.* Mcfarland & Co., Jefferson, North Carolina, 1996.

Klauber, Laurence M. *Rattlesnakes.* University of California Press, Berkeley (abr. edn.), 1982.

LaBarre, Weston. *They Shall Take Up Serpents.* University of Minnesota Press, Minneapolis, 1962.

Lawson, John. *History of Carolina.* London, 1714

Morton, Thomas. *New English Canaan.* Charles Greene, London, 1637.

Perceptions of Hate Group Activity in Georgia. Georgia State Advisory Committee to United States Commission on Civil Rights, 1982.

Schwarz, Berthold. "Ordeal by Serpents, Fire and Strychnine" in *Psychiatric Quarterly*, 1960.

"Stirling." *Wonderful Account of the Rattlesnake and Other Serpents of America.* London, 1805.

AUSTRALIA

Beattie, Tasman and Rogers, Nan. *Traveller Brown.* Boolarong, Brisbane, 1988.

Bertie, Charles Henry. "The Story of Vaucluse House." Paper read to the Australian Historical Society, 1938.

Blyth, K. A. "The Taipan—Truth and Treatment." The Producers' Review, Queensland Cane Growers' Council, October, 1956.

Cann, John. *Snakes Alive!* Kangaroo Press, Sydney, 1986.

Cramp, K. R. *William Charles Wentworth of Wentworth House.* D. S. Ford, Sydney, 1918.

Crowther, W. "Mr. Charles Underwood and his Antidote, with some Observations on Snake-Bite in Tasmania," from *The Medical Journal of Australia*, January 21, 1956.

Flynn, Rev. John. *The Bushman's Companion: a Handful of Hints for Outbackers.* Brown, Prior & Co., Melbourne, 1910.

Hughes, Robert. *The Fatal Shore.* Harvill, London, 1987.

Huxley, J. (ed). *T. H. Huxley's Diary of the Voyage of H.M.S. Rattlesnake.* Chatto & Windus, London, 1935.

Jones, D. *Trinity Phoenix: A History of Cairns and District.* Printed by *Cairns Post*, Cairns, 1976.

Jones, P. *Search for the Taipan.* Angus & Robertson, Sydney, 1977.

Krefft, Gerard. *The Snakes of Australia.* Sydney, 1869.

Lang, John. *Remarkable Convicts. Or Recollections of Botany Bay.* Ward & Lock, London, 1860.

MacGillivray, J. *Narrative of the Voyage of H.M.S. Rattlesnake,* two vols. London, 1852.

Marjoribanks, Alexander. *Travels in New South Wales.* Smith, Elder & Co., London, 1847.

Masci, Paul and Kendall, Phillip. *The Taipan.* Kangaroo Press, Sydney, 1995.

Mirtschin, P. and Davis, R. *Dangerous Snakes of Australia.* Rigby, Adelaide, 1982.

Morris, Jan. *Sydney.* Penguin, London, 1992.

Mourot, S. *Vaucluse House: Origin of this Historical Monument.* Private pamphlet, 1950.

Pelham Papers. British Library ms. no. 33107.

Shine, Richard. *Australian Snakes: A Natural History.* Reed Books, Sydney, 1991.

Sutherland, Struan. *Australian Animal Toxins.* Oxford University Press, Melbourne, 1983.

Wheelwright, Horace. *Bush Wanderings of a Naturalist: Or Notes on the Field Sports and Fauna of Australia Felix.* Routledge, Warne & Routledge, London, 1861.

Williams, David. *Observations on the Australian Taipan* and *The Death of Kevin Budden.* Unpublished ms.

Worrell, Eric. *Song of the Snake.* Angus & Robertson, Sydney, 1958.

AFRICA

Bingham, Elijah. "Snake Stories from Africa." Missionary pamphlet, 1968.

Callaway, Reverend H. *The Religious System of the Amazulu.* Trübner & Co., London, 1870.

Carnochan, F. G. and Adamson, H. C. *The Empire of the Snakes.* Hutchinson, London, 1935.

Coombs, Mary and Dunachie, Susie. *Nyoka 94.* Oxford University Exploration Club to Kenya (Unpublished ms.).

Fitzgerald, William Walter Augustine: *Travels in the Coastlands of British East Africa,* Chapman & Hall, London, 1898.

Fitzsimons, F. W. *Snakes.* Hutchinson & Co., London, 1932.

Green, Lawrence G. *Secret Africa.* Stanley Paul, London, 1936.

Hardy, Ronald. *The Iron Snake.* Collins, London, 1965.

Hobley, C. W. *Bantu Belief and Magic.* Frank Cass, London, 1967 (2nd edn.).

Ionides, C. J. P. *Mambas and Maneaters.* Holt, Rinehart & Wilson, New York, 1966.

Lane, Margaret. *Life with Ionides.* Hamish Hamilton, London, 1963.

Loveridge, Arthur. "Snakes and Snake-Bite in East Africa" in *Bulletin of the Antivenim Institute of America,* January 1928, pp. 106–117.

Marais, Eugene. *The Road to Waterberg.* Human & Rousseau, Cape Town, 1972.

Meinertzhagen, Richard. *Kenya Diary 1902–1906*. Oliver & Boyd, Edinburgh, 1957.

Patterson, J. H. *The Maneaters of Tsavo*. Macmillan, London, 1907.

Pitman, Charles R. S. *The Snakes of Uganda*. Uganda Society, Kampala, 1938.

Price, Willard. *Gorilla Adventure*. Jonathan Cape, London, 1969.

Shircore, Dr. J. O. "Two Notes on the Crowing Crested Cobra" in *African Affairs*, Journal of the Royal African Society, October 1944.

Spawls, Stephen. *Sun, Sand and Snakes*. Harvill Press, London, 1979.

Thomson, Joseph. *Through Masai Land*. Sampson Low, London, 1885.

Watson, Lyall. *The Lightning Bird*. Hodder & Stoughton, London, 1982.

Willoughby, W. C. *The Soul of the Bantu*. Doubleday, New York, 1928.

Wykes, Alan. *Snake Man*. Hamish Hamilton, London, 1960.

INDIA

Crooke, W. *Popular Folklore and Religion of Northern India*. Constable & Co., London, 1896.

Daniel, J. C. *The Book of Indian Reptiles*. Bombay Natural History Society, 1983.

Deoras, P. J. *Snakes of India*. National Book Trust, Delhi, 1965.

Dubois, Abbé J. A. *Hindu Manners, Customs and Ceremonies*. Translated from the French by Henry K. Beauchamp, Clarendon Press, Oxford, 1897.

Durham, W. S. *Snake Worship and Other Topics*. Bombay, 1948.

Fayrer, Sir Joseph. *Thanatophidia of India*. J. & A. Churchill, London, 1872.

Mandlik, Vishwanath. "Serpent Worship in India: The Nag Panchami Holiday as it is now Observed" in *Speeches and Writings*. Native Opinion Press, Bombay, 1896.

Miller, Harry. "The Cobra, India's Good Snake" in *National Geographic*, 1970.

Rivett-Carnac, J. H. "The Snake Symbol in India" in *Journal of the Asiatic Society of Bengal*, 1879.

Russell, Patrick. *An Account of Indian Snakes Collected on the Coast of Coromandel*. London, 1796.

Sinclair, Gordon. *Foot-Loose in India*. Farrar & Rinehart, New York, 1933.

Vogel, J. Ph. *Indian Serpent-Lore or the Nagas in Hindu Legend and Art*. Arthur Probsthain, London, 1926.

Wall, Col. F. *The Poisonous Terrestrial Snakes of our British Indian Dominions*. Bombay Natural History Society, 1908.

Whitaker, Romulus. *Common Indian Snakes; A Field Guide*. Macmillan, Madras, 1978.

INDEX